Engineering Magnetic, Dielectric and Microwave Properties of Ceramics and Alloys

Edited by

Charanjeet Singh

Department of Electronics and Communication
School of Electrical and Electronics Engineering
Lovely Professional University
Phagwara, Punjab
India

Published by **Materials Research Forum LLC**
Millersville, PA 17551, USA

Published as part of the book series
Materials Research Foundations
Volume 57 (2019)
ISSN 2471-8890 (Print)
ISSN 2471-8904 (Online)

Print ISBN 978-1-64490-038-3
eBook ISBN 978-1-64490-039-0

Distributed worldwide by

Materials Research Forum LLC
105 Springdale Lane
Millersville, PA 17551
USA
http://www.mrforum.com

Manufactured in the United States of America
10 9 8 7 6 5 4 3 2 1

Table of Contents

Preface

In the current era of technological applications, the exploration in ceramics viz-a-viz dielectric/magnetic/magnetodielectric have been pivotal for the realization of domestic, industrial, military and medical applications. The investigation associated with tunable properties of ceramics through material science aspect (synthesis procedure, doping) and engineering component (microwave properties in the near and far field, impedance matching) have translated their possible use to reach every corner of the world at an affordable price. The present volume entitled "Engineering Magnetic, Dielectric and Microwave Properties of Recent Ceramics", to be published as part of the book series "Materials Research Foundation" by Materials Research Forum LLC, USA contains a total of eight chapters. The topics are focused on biomaterials, ferrites, alloys, capacitors, multiferroics, microwave absorbers and perovskites materials. The authors have tried to cover basic concepts, synthesis procedures, and different applications of magnetic ceramics, biomaterials, multiferroics, perovskite materials underlying the pertinent mechanisms and models.

The first chapter by M.P.F. Graça *et al.* is focused on different physical and biological properties of bioceramics for orthopedic applications. The authors have mentioned the pre-requisite of biomaterials for body implantation, classification of biomaterials, a brief review of different reports associated with biomaterials. The main discussion revolves around two kinds of biomaterials; Hydroxyapatite and Bioglass®. The different reaction mechanisms at the interface of implant/physiological media interface along with quaternary systems, ternary systems, binary systems of bioactive glass systems have also been explored.

The second chapter by S.A. Palomares-Sánchez *et al.* presents the review of the preparation methods, properties and applications of $PbFe_{12}O_{19}$ hexaferrites. The authors have described crystal structure, structural parameters and preparation methods such as ceramic method, sol-gel method, self-propagating high-temperature method, citrate precursor method of lead ferrites etc. The coercivity, saturation magnetization, anisotropy field etc. of lead ferrite and their thin films were also discussed. The different objectives of reported Co-Sn-Ho, La, Ga, Co-Ti etc. doped lead ferrites were also relisted.

The third chapter contributed by Sergie V. Trukhanov *et al.* involves investigation on Ni-Fe alloys which are used to protect/shield the electronic devices from external magnetostatic fields. Authors have prepared $Ni_{1-x}Fe_x$ ($0<x<0.5$) films and $Ni_{80}Fe_{20}/Cu$ multilayered structures by electrodeposition method. The maximum magnetic permeability, coercive force, saturation field and shield effectiveness of the prepared compositions with different thickness have been measured.

The fourth chapter by Dr. Maciej Jaroszweski *et al.* is focused on the development of capacitor materials technology. The different properties of dielectrics viz-a-viz electric field, polarization, dipole moment and categorization of the ceramic capacitor have been explained. The manufacturing process of the disk capacitor and multi-layer ceramic capacitor (MLCC) has been discussed. The theoretical concepts of super-capacitor along with its' charge and discharging have been mentioned. The performance of different materials for electrodes of the supercapacitor has also been listed.

In the fifth chapter, I.A. Abdel-latif have reviewed the role of multiferroics in spintronics field. After the discussion about ferroelectric and ferromagnetism along with basic prerequisites, author has explained about the potential use of magnetoresistive random access memory and its advantage over static and dynamic random access memory. The comparison has been done between spin valve transistor and conventional transistor. The hurdles are mentioned for practical implementation for spin valve transistor and gist of associated reports for the same transistor has been also prepared.

In the sixth chapter, Hesham Zaki reports the dielectric properties of Spinel ferrites $Li_{0.5+0.5x}Ge_xFe_{2.5-1.5x}$, $Cu_{1+x}Ge_xFe_{2-2x}O_4$, $Cu_{1+x}TixFe_{2-2x}O_4$, $Cu_xFe_{3-x}O_{4+\delta}$ has been presented. The author has discussed the effect of doping, temperature and frequency on DC and AC conductivity, dielectric constant and dielectric loss and dielectric loss tangent. The necessary mechanism and models behind their variation along with mathematical equations is also described.

In the seventh chapter, K. Sakthipandi *et al.* have discussed structural and magnetic properties of rare earth doped manganite perovskites. The effect of ionic size substitution on perovskite structure, ED and SAED pattern of various compounds have been studied. The magnetic properties such as magnetoresistance, magnetic ordering, coercivity, saturation magnetization etc. are also covered.

The eighth chapter by Charanjeet Singh *et al.* is focused on the design of microwave absorbers and absorption characteristics of ceramics. The different types of dielectric relaxation have been reported with the help of Cole-Cole plots. The relationship between hysteresis parameters and microwave absorption has been elaborated mathematically. The authors presented the effect of eddy current loss, magnetic and dielectric loss on absorption. The role of quarter wavelength mechanism and impedance matching based on input impedance absorber as well as difference of the loss tangent of absorber in optimization of microwave absorption are mentioned.

The editor expresses heartfelt thanks to Thomas Wohlbier, Materials Research Forum, LLC, USA for giving the opportunity to debut the footsteps in the publishing world with a special volume in the form of a book. Further, the editor is thankful to Dr. R.B Jotania and Dr. Sukhleen Bindra Narang for providing critical comments during this work.

Charanjeet Singh,
Professor
Department of Electronics and Communication
School of Electrical and Electronics Engineering
Lovely Professional University
Phagwara, Punjab
India

Eng. Magnetic, Dielectric and Microwave Properties of Ceramics and Alloys   Materials Research Forum LLC
Materials Research Foundations **57** (2019) 1-22   doi: https://doi.org/10.21741/9781644900390-1

Chapter 1

Physical and Biological Properties of Biomaterials Intended for Bone Repair Applications

S.R. Gavinho[1]*, M.P.F. Graça[1]*, P.R. Prezas[1], C.C. Silva[1,2], F.N. Freire[3], A.F. Almeida[3], A.S.B. Sombra[3]

[1] I3N and Physics Departement, Aveiro University, aveiro, Portugal

[2] Federal University of Maranhão, Imperatriz, Brasil

[3] Federal University of Ceará, Fortaleza, Brasil

silviagavinho@ua.pt; mpfg@ua.pt

Abstract

Hydroxyapatite is a biomaterial which has attracted a great deal of attention because of its chemical similarity with the composition of the mineral phase of human bones – biologic hydroxyapatite. Among its most recurring applications are coatings for orthopedic and dental implants, maxillofacial surgery, otolaryngology, scaffolds for bone tissue engineering and application as powders in total hip and knee surgeries. On the other hand, bioactive glasses belonging to the system SiO_2-P_2O_5-CaO-Na_2O are reported to stimulate host bone regeneration at a higher rate than any other known biomaterial. In the present contribution, the essential and relevant physical and biological properties of these biomaterials will be discussed.

Keywords

Bioceramics, Hydroxyapatite, Mechanical Alloying, *In vivo* Bioactivity, Bioglass[®]

Contents

1.   Introduction

The primary function of biomaterials intended for orthopedic interventions is not only filling a gap caused by the loss of hard tissue but also to provide physical and biological compatible characteristics, i.e., they must be biocompatible with the living tissue of the individuals who will receive the implant or graft [1, 2].On the other hand, biomaterials have several important characteristics that should be considered for implantation, such as cost of fabrication, easiness of manufacturing, biocompatibility, non-carcinogenic, appropriate density, mechanical stability, ideal weight, chemically functionality, reproducibility, non-toxicity, and stimulation of biological reactions [3]. The stimulation of favorable biological reactions is a very important property known as bioactivity, i.e., the implant has the ability to form a strong bond with the biological host bone, by stimulating the cellular activity. Other important properties of these biomaterials are the bioresorbability, i.e., the withdrawal process of the material may be through dissolution, and the osteoconductivity, i.e., the implant has the ability to act as a template to guide the formation of the newly forming host bone [2].

Biomaterials can be classified according to their origin, physiological behavior and chemical nature [4]. As for its origin, biomaterials can be autologous, allogeneic, and alloplastic xenogenous. When the material is to be deployed from other individuals within the same species, it is known as a halogen. These may be obtained from cadavers, which subsequently are processed and leveled. While they are feasible to be implemented, this process presents risks of contamination and rejection at the time of deployment [5, 6].

According to the interaction with the tissue and the physiological environment, biomaterials can be divided into bioinert, bioactive, resorbable and tolerable. The bioinert materials are those that are accepted by the body, however, there is virtually no release of any component, and they do not react with the surrounding tissue. Instead, they are encapsulated by fibrous tissue and separated from the environment [6]. Nonetheless, in 1986, at the Conference of the European Society for Biomaterials, in England, it was determined that the term bioinert should not be used because all materials lead to some type of tissue response from the host. However, the term bioinert is still common in descriptions of biomaterials[2]. Alumina (α - Al_2O_3), zirconia (ZrO_2) and titanium dioxide (TiO_2) ceramics are examples of bioinert biomaterials[7].

As for bioactive materials, they allow osteointegration of bone tissue by chemical bonding in the absence of the fibrous casing characteristic of bioinert materials. They also allow osteoconduction of the growing host bone, by promoting a biological response at the interface that is related to the bonding between the hard tissue and the implanted material [8].

In the in-vitro test, using simulated body fluid (SBF), bioactive materials "lose" sodium ions and form a surface film rich in SiO_2, followed by the formation of a layer of calcium phosphate gel which is initially amorphous and that gradually develops into a layer of polycrystalline clusters of hydroxycarbonate apatite (HCA). The formation of the HCA layer is important for the process of dissolution of the glass and enables physical-chemical processes that leading to the formation of chemical bonds between the glass surface and the formed bone tissue in the surgical intervention region. The specific surface activity characterized by this interfacial ion exchange between the biomaterial particles and SBF or the fluid of the neighbouring tissues is the most critical factor that determines bioactivity.

Bioactive materials are classified into: osteoinductive and osteoconductive. The implants that generate an intracellular and extracellular response at the biomaterial-tissue interface is designated as osteoinductive: they promote the development of a biocompatible implant site surface, promoting the formation of bone cells. Hydroxyapatite, HA, is an example belonging to this group of biomaterials [9].

According to their chemical nature, biomaterials can be classified as either natural or synthetic. Pure collagen, or mixed with other natural compounds, and bones obtained from persons, or another animal source (such as bovine bone), are considered natural biomaterials. On its turn, synthetic biomaterials include metallic, ceramic, polymeric and composite materials [4].

Bioceramics based on calcium phosphates, most notably HA, have been widely used in medicine and dentistry in the last 20 years since they present the possibility of repair and/or replace parts of the human body with pre-fabricated materials. Among their numerous applications, they are applied as films to coat metallic orthopedic and dental implants, in maxillofacial surgeries and otolaryngology [10]. Regarding their use as films on metallic implants, they promote adhesion between the prosthesis and the host bone [11]. HA is present in vertebrates by composing the mineral phase of our bones, representing 65 to 70 % in weight of human bones [12]. In fact, human bones are composites formed by an organic and a mineral phase. In the organic phase, the fibers of collagen serve as a matrix for the precipitation of HA (mineral phase), determining the structures and alignment of the crystals.

Nowadays, there is no way to imagine modern medicine without the use of metallic materials, polymers, vitreous carbon, Bioglass® and bioceramics in implants. Research on these new materials has increased in recent years due to their enormous importance in trying to achieve implants with highly reduced rejection probability by the human body. The possibility of anatomical and functional construction of human body parts has led to several areas of science seeking new alternatives to the materials already existing. Therefore, the study and development of these types of materials are of fundamental importance for the reconstruction and/or replacement of parts of bone and cartilage of the human body by specific forms called implants, prostheses or prosthetics devices [4, 13]. Biomaterials, as implants in the human body or in animals [14, 15] can be additionally classified as permanent, as in the case of pacemakers, bone and dental implants, or temporary, as it is the case for the catheter. Their mechanical and chemical properties, in a general sense, and the degree of acceptance by the "host", in the case of the human body, determine their biological performance or biocompatibility.

The fundamental properties required for this type of materials are the complete biocompatibility allied with the highest possible mechanical resistance, efficient bioactivity, chemical resistance, especially against blood fluids and a good mechanical (articulation) performance. More specifically, biocompatibility is understood as the compatibility between the material of the prosthesis and the biological system and can be classified as structural compatibility or superficial. Structural compatibility refers to the mechanical adaptation of the implant with the environment surrounding the site of implantation. As an example, between the implant and the host bone, the transmission of mechanical forces and stimuli between them plays a vital role in bone remodeling. Surface biocompatibility is another fundamental property in this case since the surface of the prosthesis is directly in contact with the physiological media. This surface compatibility has a higher degree of efficiency when using bioactive materials on the surface of the prosthesis, as it is the case of HA and Bioglass® films [16].

HA has a similar structure to the mineral phase of human bones, providing its bioactive properties. After some time of implantation, there is a growth of new host bone tissue over the film of HA. In the work of Vogel and Höland, one of the tests for the biocompatibility of vitro-ceramics was performed by subjecting them to cell culture, and after some time they noticed the absence of evidence of deterioration. To evaluate the bioactivity, Vogel and Höland performed implantation of vitro-ceramic cubes in guinea pigs, in the region of the tibia [14]. After twelve weeks, these implants were removed.

The force required to remove them was in the magnitude of $5N/mm^2$, which is eight times greater than the force reported for other implants.

It is well known that, in the chemical constitution of bones and teeth, we have the presence of calcium and phosphorus elements [17]. It is also known that these elements are found in nature as fluoroapatite ($Ca_{10}(PO_4)_6F_2$), belonging to the series of isomorphic minerals known as Apatites [18].

As it was said, the components of bones can be classified into organic and inorganic. The organic component, comprising about 25 to 30%, by weight, is composed primarily of collagen molecules. The inorganic component consists of an amorphous and a crystalline part. The first amorphous phase appears predominantly in young bones, being later partially transformed in the crystalline phase HA [19], with chemical formula Ca_{10} $(PO_4)_6(OH)_2$ and a well-defined crystallographic structure [20].

The first generation of implants was made using materials, such as some metals, that are known to cause inflammation and rejection from the body. The second generation of biocompatible materials was composed mainly of inert materials that do not biodegrade and have no significant influence on metabolism, i.e., they do not possess bioactive properties. The third generation of materials was used as an inducer of metabolic processes and growth factor of hard tissues. The materials of this generation are known as bioactive or metabolic inducers. Finally, the fourth generation is denominated as cellular composites, used in the transport system for cell proliferation and differentiation. Several authors cite examples of biomaterials composed of metals and alloys [21], organic polymers [13], ceramics [22, 23] and Bioglass® [23].

Combinations of polymers, composites made of polymers with carbon fibers and bioactive ceramics have also been used in the construction of prostheses. There is also a need to use different types of materials in the production of a single prosthesis, such as titanium alloys to form the prosthesis body (and with biocorrosion resistance), a thin HA or bioglass® film to allow a strong bond between the host bone and the prosthesis (bioactivity) and a spherical surface of Alumina at the end of the prosthesis to improve its strength.

Due to its properties, mainly its biocompatibility, HA has become one of the most studied substances in the last twelve years [19, 24-28]. As referred, one of the most interesting properties of HA is its bioactivity. This property is directly related to its porosity. Porosity studies in HA implants in dogs [24] and in the human jaw [28] showed that there was bonding with the new host bone inside the pores accessible to the physiological fluid and bone specialized cells, known as osteoclasts and osteoblasts. In the case of implants with dense HA, studies in dogs [25] and biopsies performed in humans [26] showed that

Eng. Magnetic, Dielectric and Microwave Properties of Ceramics and Alloys   Materials Research Forum LLC
Materials Research Foundations **57** (2019) 1-22   doi: https://doi.org/10.21741/9781644900390-1

these implants were involved by human tissue fibers, with a variable formation of new bone tissue. Despite these qualities, the mechanical properties of HA are relatively weak. Although it withstands compressive forces in the order of 250 MPa, its resistance to fatigue is still very low [29]. One of the solutions to this problem is to use metallic prostheses covered with HA (or other bioactive material) films [15, 30] combining the mechanical properties of the metallic component with the bioactive properties of the coating. There are several benefits arising from the use of HA layers: the rapid adaptation of the implant with the tissue around it, reducing the restoration time [31], the increase in the host bone formation rate [32], optimal fixation between the implant and the bone [33] and the reduction in the release of metallic ions [34]. Despite this, special care must be taken regarding the thickness of the film, since thin films can degrade quickly over time. Experiences have shown that a few weeks after implantation, about 10% of the apatite surface dissolves during the process of interaction with the physiological media. On the other hand, films with 100 up to 150 mm may be compromised due to fatigue, and delamination can occur. The most suitable thickness that has been found is around 50 mm [29]. In addition to ceramics, other materials have shown great potential as substitutes/alternatives of HA films, such as Bioglass® [14, 15] or vitro-ceramics bioactive materials. One of the most significant scholars of this type of materials was Larry L. Hench [27], who developed the Bioglass® formulations. The main advantages of the Bioglass® include their excellent bioactivity and osteointegration ability, high flexibility of composition and chemical stability. Actually, bioactive glasses belonging to the system SiO_2-P_2O_5-CaO-Na_2O (Bioglass®) are reported to be able to stimulate more host bone regeneration than any other known biomaterial. They have excellent biocompatibility, not interfering with the tissue cells (osteoclasts and osteoblasts) normal behavior. Some examples of bioglasses include compositions such as those developed by Ogino and Hench [35], that studied films of SiO_2-Na_2O-CaO-P_2O_5, SiO_2-Na_2O, SiO_2-Na_2O/K_2O-MgO-Al_2O_3-F and SiO_2-Na_2O-CaO-Al_2O_3-B_2O_3, among others [14, 15]. With appropriate heat treatment, Bioglass® films can also improve their bioactivity, allowing a stronger bond with the host bone tissue [36]. Several processes [37, 38] are used to synthesize HA, intended for several types of applications, including, of course, a temporary substitute for human bones [38-41]. The hydrothermal synthesis is characterized by a reaction in aqueous solution in a closed container, with temperature and pressure as the process variables. A specific hydrothermal synthesis method consists in submitting aqueous solutions containing Ca^{+2} and PO_4^{-3} ions to high temperatures (200-500°C).

Recently, considerable interest has been focused on the use of biodegradable or bioabsorbable materials such as collagen [42, 43], gelatin [44] and synthetic polymers such as polyhema[45]. Collagen, which exists in a variety of morphological forms [46], is the most abundant protein structure in connective tissue, in addition to having a long history as biomaterial [47]. In the bone structure, individual collagen molecules, in which the rods are semi-flexible and have a diameter of about 1 nm and a length of approximately 280 nm, undergo self-regrouping to form complex network structures.

Some biological materials and biopolymers exhibit a uniaxial polar orientation of dipole molecules and can be considered bioelectrodes. Polymeric materials and biocompatible polymers are now used extensively in conjunction with an electrical polarization treatment suitable for biomedical applications such as anti-biogenetic surfaces and membranes artifacts [48]. Studies of pyro- and piezoelectricity in various types of biological systems showed the presence of a natural polarity in the structure of several parts of animals and plants. In many natural structures of polar molecules like proteins, they are aligned parallel with a preferred direction according to the polar axis, to form a crystalline structure. These structures can be seen as natural electrodes, because an intrinsic electrical polarization, pyroelectricity, and piezoelectricity can be detected along the axial direction [49]. The piezoelectric properties of collagen have been investigated in natural components such as bone and tendons, in order to try to access the role of such properties in the normal bone's growth [48].

1.1. Bioactive glasses

Researchers have been developing new materials with high biocompatibility, durability and a high-performance for applications in medical devices. In addition, another fundamental characteristic of material useful for tissues engineering is the bioactivity, i.e., the ability for promoting a strong bond with the biologic host hard tissue, and consequently, aiding in their repair. Until now, several materials, such as metals, polymers, glass-ceramics, have been applied in this field. In the past 20-30 years, bioactive glasses and glass-ceramics have been gathering a considerable amount of interest among the scientific community [50].

As an example, in the past decades, metallic materials have been extensively used as dental implants due to their advantageous mechanical properties. The most commonly used metals are usually titanium or titanium alloys and, cobalt alloys. However, there some concerns related to the utilization of these materials, especially in what concerns the release of metallic ions to the physiological media, which can lead to hypersensitivity reactions to the metal. Nevertheless, titanium implants are most commonly used for *in vivo* applications because they have low density, high biocorrosion resistance, and excellent mechanical strength. Despite these properties, a higher level of interaction of the implant with the physiological environment is required, in order to promote a stronger bond between the implant and the host tissue, i.e., an efficient osseointegration. Thus, bioactive biomaterials are an indispensable tool for developingthe stability and clinical success of the metallic implants, by stimulating fast bone regeneration. This osseointegration is improved by coating the metallic implant surface with bioactive biomaterials. For such purpose, Bioglass® is one of the most extensively studiedmaterials [51].

Bioactive glasses (such as the Bioglass®), are silicate-based amorphous materials that, in a biological system, are capable to stimulate the cellular activity at the genetic level, being more effective in promoting the host bone regeneration than any other known bioactive material, including ceramics such as hydroxyapatite (Hap) [52]. Additionally, bioglasses are also reported to be able to enhance revascularization, osteoblast adhesion, differentiation and proliferation, enzyme activity and differentiation of mesenchymal stem cells, as well as osteoprogenitor cells [53].

The biocompatibility of bioglasses is directly related to their chemical composition, integrating minerals that are present in the body, and the molecular proportions of the calcium and phosphorous oxides are similar to the characteristic of bones. The surface of a bioglass-coated metallic implant, when reacting with the physiological media, starts a dissolution process and is quickly covered with a silica$-$CaO/P_2O_5-rich gel layer that, subsequently, mineralizes into carbonated Hap. For higher dissolution rates, it is to be expected a faster host bone regeneration. The deposited gel layer is structurally similar to the Hap matrix, and consequently, induces osteoblasticadhesion and differentiation and new hostbone deposition [54].

The original Bioglass® formulation (the 45S5 formulation) was discovered by Hench et. al. Its composition is 45% SiO_2, 24.5% Na_2O, 24.5% CaO, 6% P_2O_5 (molar %)[55]. The 45S5 formulation exhibits the highest bioactivity index of any known biomaterial intended for bone/teeth repair and augmentation and is still recognized as the gold standard of bioactive materials [56]. Nevertheless, many different compositions have been studied, which can differ in the type of oxide added, as well as in the molar percentages of the components, allowing to obtain a material that can be bioinert, bioresorbable or bioregenerative. In addition to the composition, the synthesis method can also influence the biological properties of the bioactive glasses and glass-ceramics [50].

Eng. Magnetic, Dielectric and Microwave Properties of Ceramics and Alloys   Materials Research Forum LLC
Materials Research Foundations **57** (2019) 1-22   doi: https://doi.org/10.21741/9781644900390-1

2.   Reaction mechanism

In order to understand the importance of the composition and the molar percentages, it is necessary to understand the reaction mechanisms at the implant/physiological media interface.

Considering the Hench classification system describing the bioactivity of material, a new bi-class classification system was proposed. Class A materials (e.g., Bioglass®) are osteoproductive, i.e., their bioactive surface allows osteogenic stem cells colonization and differentiation, stimulating both intracellular and extracellular responses. On the other hand, Class B materials are osteoconductive, stimulating the bone migration and acting as a template to guide the host bone formation. These materialsact only at the extracellular level of the target tissue (e.g., synthetic Hap) [57].

As Fig.1 shows, keeping the P_2O_5molar composition fixed, several different reactions with the physiological media can be obtained, by changing the other elements molar composition.

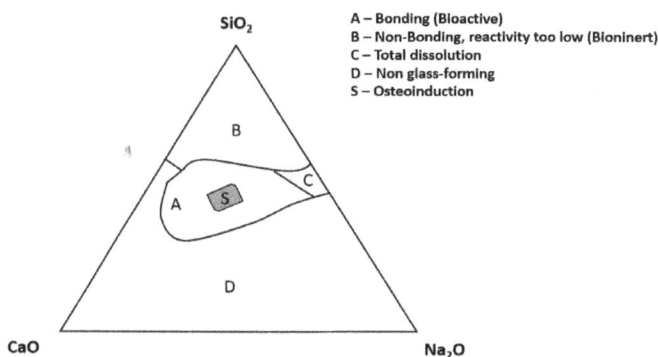

Figure 1 –*Property changes of bioglass materials depending on their composition (adapted from[57]).*

The reaction that takes place on the surface of the bioactive material, forming the carbonated Hap (HCA) layer and the subsequent bonding of the implant with the host tissue, occurs in five main stages [58]. In the first stage, happens the formation of an amorphous layer of calcium phosphate, which crystallizes due to the nucleation and growth of a phase similar to apatite, giving rise to nanocrystals of HCA, identical to those that compose the mineral phase of our bones. When these nanocrystals interconnect with the collagen fibers, they form a layer that binds the bioactive material to the host living

9

Eng. Magnetic, Dielectric and Microwave Properties of Ceramics and Alloys   Materials Research Forum LLC
Materials Research Foundations **57** (2019) 1-22   doi: https://doi.org/10.21741/9781644900390-1

tissue [59]. The interaction of the bioactive glass at the cellular level is a very complex process. However, Hench et al. proposed a possible mechanism that explains the formation of the HCA layer. In this mechanism, when the bioglass comes in contact with the physiologic fluid, a rapid exchange of ions occurs, promoting the dissolution of the network. This dissolution is related to the loss of soluble silica, such as $Si(OH)_4$ (breakage of Si-O-Si bonds), forming Si-OH bonds (silanols groups) on the bioglass surface. Subsequently, the condensation and repolymerization of a SiO_2-rich layer on the surface of the bioglass occur, and an amorphous CaO-P_2O_5-rich film is coupled to this layer due to the Ca^{2+} and PO_4^{3-} ions migration. Finally, the crystallization of this amorphous film occurs due to the incorporation of OH^-, CO_3^{2-} or F^- anions present in the physiological media, forming an HCA or hydroxy fluorapatite (HCFA) layer [58].

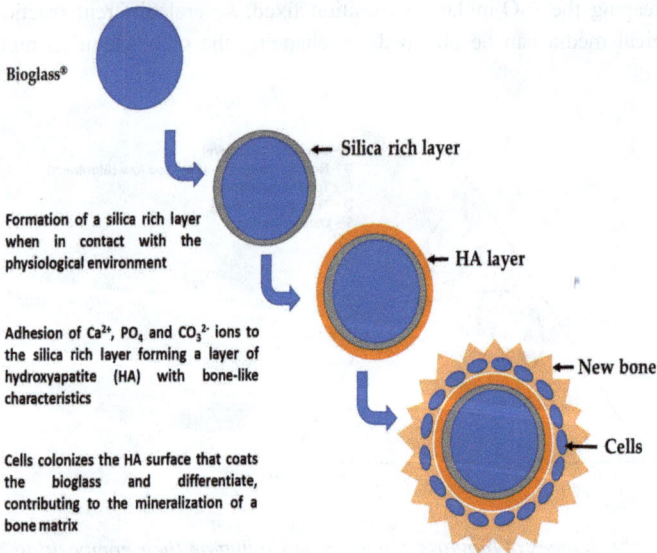

Bioglass®

Formation of a silica rich layer when in contact with the physiological environment

← Silica rich layer

Adhesion of Ca^{2+}, PO_4 and CO_3^{2-} ions to the silica rich layer forming a layer of hydroxyapatite (HA) with bone-like characteristics

← HA layer

Cells colonizes the HA surface that coats the bioglass and differentiate, contributing to the mineralization of a bone matrix

← New bone

← Cells

Figure 2 *– Representative scheme of the Bioglass®surface/physiological media reaction (adapted from [60]).*

3. Bioactive glasses systems

3.1 Quaternary systems

In the field of tissue engineering, one cannot forget Hench's contribution. About the bioglass, he reported that, in particular, compositions with SiO_2-P_2O_5-CaO-Na_2O systems form a powerful bond with the bone [60].

The original 45S5 Bioglass® features a quaternary system composition (SiO_2-Na_2O-CaO-P_2O_5). It is a silicate glass, and its low molar percentage of SiO_2 (glass former component), its high concentration of Na_2O and CaO (glass network modifiers, fundamental to stimulate the host bone regeneration) and its high CaO/P_2O_5 molar ratio are responsible for its bioactivity (compared to other silicate bioglasses with higher durability) [61]. The low silica content and the presence of sodium and calcium ions in the glass network are responsible for the fast bioglass/biological environment surface dissolution reaction. Compared to other biomaterials, the 45S5 bioglass binds faster to the host bone, and *in vitro* studies indicate that the products of the dissolution reaction at the bioglass interface stimulate the osteoprogenitor cells activity at the genetic level. It is challenging to produce porous bioactive glass scaffolds, due to limitations related to the sintering heat treatments. These restrictions, are mainly associated with the crystallization of the vitreous matrix: the interval between the glass transition temperature (T_g) and the onset of crystallization is small, making difficult to sinter dense networks, making the scaffolds fragile. Researchers are trying to overcome this problem by changing the composition of the matrix, in order to improve the resistance against [60, 62]. The long-term effects of SiO_2 content *in vivo* are unknown because the dissolution reaction is sometimes incomplete, with some SiO_2 remaining in the scaffold [61]. This issue can be avoided using the sol-gel synthesis process, where the silica network is developed at room temperature. In addition to changes in the preparation method, ionic modification in the vitreous network is an important parameter capable of improving the properties of bioglass, which can accelerate hard tissue regeneration, increase antibacterial activity and angiogenesis [63].

In the following section, several synthesis methods and different bioglass compositions were compared and modified, in order to optimize the bioglass characteristics.

The 46S6 bioglass formulation (46% SiO_2, 24% CaO, 24% Na_2O, 6% P_2O_5,wt%) has been synthesized by various methods, namely, sol-gel and melt-quenching. The addition of some compounds, such as K_2O and Na_2O, used in the melt-quenching method, allows to reduce the melt temperature and makes the materials more soluble in aqueous media. This feature is crucial for the biomaterial surface/physiological media interaction. However, the 46S6 formulation containing high amounts of Na_2O has not yet been

11

Eng. Magnetic, Dielectric and Microwave Properties of Ceramics and Alloys   Materials Research Forum LLC
Materials Research Foundations **57** (2019) 1-22   doi: https://doi.org/10.21741/9781644900390-1

achieved nor optimized. The sol-gel-derived bioglass particles have a relatively large size because of the long gelation and aging times. However, researchers have successively modified the sol-gel method in order to reduce the required gelation time, by applying an acid-base reaction that led to particles with smaller sizes (40-60 nm). The smaller size of the particles promotes a higher cellular activity, due to the higher reactivity of the particles, because they have a larger surface area/volume ratio. In fact, it is reported that the deposition of the apatite layer on the glass prepared by sol-gel is faster compared to the melt-quenching method (figure 3). Therefore, sol-gel synthesized 46S6 glasses are confirmed to be bioactive and to be potential materials for bone replacement and repair applications [64].

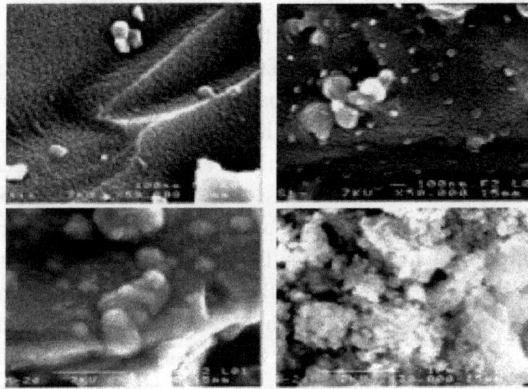

Figure 3 –*SEM micrographs; On the left: 46S6 bioglass nanoparticles synthesized by melt-quenching before (above) and after (below) immersion in Simulated Body Fluid (SBF) for 2 days. On the right: Bioglass nanoparticles synthetized by sol-gel method before (above) and after (below) immersion in Simulated Body Fluid (SBF) for 2 days [64].*

Besides the 46S6 glass, researchers have developed other quaternary systems. The bioactivity of the 52S4 formulation (52% SiO_2, 30% CaO, 14% Na_2O, 4% P_2O_5,wt%) was evaluated in order to access the influence of the synthesis method. Both methods, sol-gel, and melt-quenching produce bioactive glasses, yet, the bioactivity of the sol-gel synthesized glasses is enhanced by the presence of porosity in the material, increasing the specific surface area, and thereby the ionic exchanges between the material and the biological medium. When 52S4 glasses, synthesized by both methods, were immersed in SBF (Simulated Body Fluid), the bone-like apatite layer was formed earlier in the bioglass prepared by sol-gel, due to an increased reabsorption rate [65]. Once again, the synthesis method was important to acquire the optimal characteristics for its intended application, being the sol-gel method more favorable in terms of structural characteristics and bioactivity level.

Regarding the properties dependence on the bioglasses compositionchange, a study is presented comparing two bioglasses with different silica molar percentages. Indeed, researchers have evaluated the behavior the S53P4 (53% SiO_2, 23% Na_2O, 20% CaO, 4% P_2O_5, wt%). This bioglass has been increasingly applied in clinical practice, especially in grafts and osteomyelitis treatment. It has the ability to stimulate hostbone formation and has antibacterial properties [66]. Additionally, the performance of an S53P4 glass was compared with another with a lower amount of silica, in the treatment of dentin hypersensitivity, i.e., hypersensitivity at the mineralized tissue located between the crown enamel or root cementum and the pulp of the tooth. The S53P4 glass presented bioactive properties, promoting the formation of a dense and uniform CaP layer on the silica-rich surface and representing an efficient source of silica and calcium for the biomineralization process. The treatment with this bioglass also showed high concentrations of adsorbed silica, that increased with immersion time in the biological medium. Based on this study, it was concluded that S53P4 could be used clinically in the mineralization of dentin [67].

3.2   Ternary systems

Besides quaternary bioglasssystems, researchers have also developed and characterized ternary systems, analyzing the dependence of their properties on the compositional modifications and synthesis methods.

The 64S formulation, with a molar composition of 64% SiO_2, 31% CaO and 5% P_2O_5, is one of the ternary bioglass systems that has been researched. These glasses appeared amorphous, without showing any signal of crystallization. The bioactivity was analyzed through the osteoblastic activity when in contact with the bioglass, by evaluating one of the markers of osteoblastic proliferation and differentiation, in this case, the expression of alkaline phosphatase. The results showed that there is a high level of bioactivity in the ternary system, stimulating early differentiation of the specialized bonecells. It also showed no toxicity. In addition, SEM micrograph of the bioglass powder is shown that the grain size of bioglass is in the nanoscale range. By decreasing the molar percentage of phosphorus in the solution and increasing the dissolution of the Ca and Si ions of the bioglass in the medium, affirming the deposition of HCA, which suggests a rapid connection between the bioglass and the bone tissue [68].

The SiO_2-CaO-P_2O_5 system synthesized by the sol-gel method has been extensively studied, mainly the 55S, 58S and the 60S formulation. The high bioactivity makes them an excellent material for use in hard tissue surgery. The 58S was chosen as coating material and it has been shown that it begins to crystallize in the temperature range of 850-900°C. It was proven that the appearance of crystalline phase decreases the capacity of densification. However, it also increased its hardness. In addition, the onset of this phase makes the 58S less bioactive. The dense apatite layer is formed only when in amorphous "state", making it highly bioactive. After crystallization, the apatite layer becomes incomplete or porous [69, 70].

In addition to bioactivity, it is also beneficial that bioglasses have antibacterial properties, being one of the most important complications caused by certain biomaterials. In the case of implants, the attach used is highly susceptible to the action of bacteria, causing infection. These infections imply quite a lot of complications and later invasive treatments, being a process to avoid. Therefore, bioglass studies have been carried out, testing their action against the various type bacteria. The bacterial action is variable depending on the chemical composition of the material used. For example, the original quaternary system proposed by Hench et al. (45S5 formulation) has a high antibacterial effect when particle size reduction occurs.

In previous studies, 64S bioglass had no bactericidal effect on Escherichia coli, perhaps because it had particles in the micrometer range and the concentrations analyzed were not sufficient to have an effect on the bacteria. However, new studies suggest the hypothesis that the reduction of particle size improves the antibacterial effect of bioglass, with high concentrations of SiO_2. Three bioglass of the ternary systems were studied in aerobic bacteria in order to verify this hypothesis. The minimum bactericidal concentration (MBC) required to have the desired effect was also analyzed. The bioglass formulations used were 58S (57.72% SiO_2, 35.09% CaO, 7.1% P_2O_5), 63S (62.17% SiO_2, 28.47% CaO, 9.25% P_2O_5) and 72S (72.88% SiO_2, 17.49% CaO, 9.56% P_2O_5), and were synthesized by the sol-gel method.The prepared bioglass originated nanoparticles with sizes between 20 and 90 nm, being favorable for a better biomaterial / biological environment interaction. The nanoparticles of the 58S bioglass showed higher antibacterial activity, and its MBC for *Escherichia coli* and *Staphylococcus aureus* was 50mg/ml and for *P. aeruginosa* was 100mg/ml. The 63S formulation proved to be antibacterial when tested on the *Escherichia coli* and *Staphylococcus aureus* with an MBC of 100mg/ml, but no antibacterial effects were detected on *Salmonella typhi* and *Pseudomonas aeruginosa*. The 72S bioglass showed no toxicity. However, it did not present antibacterial characteristics. The 58S and 63S bioglasses presented antibacterial activity even in concentrations inferior to those used in the clinical scope. It was concluded that the bioglass synthesized with antibacterial properties could be used for the treatment of oral bone defects and to avoid infections in the root canal [71].

3.3   Binary systems

Simpler binary systems have also been researched. A bioactive glass was synthesized by the sol-gel method belonging to the $CaO-SiO_2$ system. In this research, the composition range of SiO_2 was varied between (50-90 mol%), allowing to study the limits of bioactivity, looking for different reactions when immersed in the biological medium. This binary system does not contain any phosphorussource, and of all the studies carried out, the most critical factor was to realize that the presence of phosphorus in the bioglass network was not necessary for the formation of the apatite phase, since the physiological media contains phosphorus naturally. These glasses are bioactive, forming an apatite layer on the surface by capturing the phosphorus from the surrounding environment. The growth rate of this layer also varies with the composition of the bioglass, being that for lower SiO_2 molar compositions and higher CaO concentrations the growth rates are higher. However, this observation may also be related to the surface area and the porosity of the bioglass [59].

Researchers discovered that the bioglass with SiO_2-CaO binary system is also bioactive as in the SiO_2-CaO-P_2O_5 ternary system when the molar concentration of SiO_2 is similar. Therefore, it would be interesting to compare the three systems. *Saravanapavan et. al.* compared the bioactivity of the S70C30 (70% SiO_2, 30% CaO, mol%) binary system, the 58S (60% SiO_2, 36% CaO, 4% P_2O_5, mol%) ternary system and the 45S5 (46.1% SiO_2, 26.9% CaO, 2.6% P_2O_5, 42.4% Na_2O, mol%) quaternary system. Comparing the texture and the particle size, S70C30 and 58S presented surface areas and pore volumes identical, varying in pore size, whereas 45S5 is denser and without pores. With respect to the dissolution profiles, the binary system bioglass initially has, in the first 2 hours, a higher Si release rate than in the 58S and 45S5 systems when in contact with the SBF. This high rate of Si release may be the result of higher Ca concentration, leading to a faster network dissolution. In the following phases, the binary bioglass behaves similarly to 45S5, and dissolution rates are comparable. It is known, as previously mentioned, that the release of silica from the bioglass (especially the 45S5) act at a genetic level, promoting osteogenesis. In addition, high Si release enhances the proliferation and differentiation of osteoblast cells. Thus, binary bioglass containing 70% SiO_2 has been shown to have comparable silicon release rates with another system. S70C30 can be used in the control of genetic expression of osteoblasts *in vivo*. The concentration of phosphorus in the SBF medium decreases for all compositions. However, this decrease becomes slower in the bioglass synthesized by melt-quenching (45S5) and in the 58S system. This is due to the existence of the P ion in quaternary and ternary systems. Regarding the rate of Ca^{2+} release, the bioglass S70C30 has a release rate twice as high than in the 58S bioglass. This may be due to the size of the pores (58S has smaller pores) [72].

Conclusions

In summary, bioactive glasses are biocompatible materials, which are generally defined as ones that elicit a specific biological response at the interface resulting in the formation of a bond between the tissues and the material. Due to this attractive property, they have been used as a filling material for the repair of bone defects, such as replacement of lost bone mass due to trauma and also as the constituent material of prosthetic devices.

A typical composition of bioactive glass is based on SiO_2, CaO, P_2O_5, Na_2O and MgO whereas other oxides such as K_2O, B_2O_3, CaF_2 may also be introduced into the system tuning the bioactive property.

At present, it is possible to develop some new biomaterials due to extensive overlaps between solid-state reactions, sol-gel chemistry and biochemistry. The sol-gel networks are excellent model systems for studying and controlling biochemical interactions within constrained matrices with well-defined textures. It is believed that enhanced bioactivity can be achieved in gel-derived materials because of their residual hydroxyl ions and micro pores, and large specific surface. Furthermore, the most recent investigations have shown that electrical polarized bioactive glasses exhibit improved bioactivity due to the induced surface charges that influence the surface activity. Based on this, it seems possible that the regulation and control of growth of the bone cells can be tuned by electrical properties of bioactive glasses.

References

[1]   Silver, F. and C. Doillon, Biocompatibility: Interactions and ImplantableMaterials., New York: VCH, (1989)

[2]   Donaruma, L.G., Definitions in biomaterials., Amsterdam: Elsevier, (1987)

[3]   Park, J., Biomaterials Science and Engineering., New York: Plenum Press, (1984)

[4]   Hench., L.L. and J. Wilson, An Introduction to Bioceramics., Singapore: World Scientific Publishing Co. Pte. Ltd., (1993). https://doi.org/10.1142/9789814317351_0001

[5]   Kakaiya, R., Miller, W.V. and Gudino, M.D., Tissue transplant-transmitted infections, Transfusion, **31** (1991) 277-284. https://doi.org/10.1046/j.1537-2995.1991.31391165182.x

Eng. Magnetic, Dielectric and Microwave Properties of Ceramics and Alloys   Materials Research Forum LLC
Materials Research Foundations **57** (2019) 1-22   doi: https://doi.org/10.21741/9781644900390-1

[6]   Stevens, M., Biomaterials for bone tissue engineering, Materials Today, **11** (2008) 18-25. https://doi.org/10.1016/S1369-7021(08)70086-5

[7]   Aoki, H., Hydroxyapatite of great promise for biomaterials,Transactions of the JWRI, **17** (1998) 107-112

[8]   Hench, L.L., et al., An Investigation of Bonding Mechanisms at the Interface of a Prosthetic Material.J. Biomed maters. Res,. **5** (1972) 117–141. https://doi.org/10.1002/jbm.820050611

[9]   Cao, W. and L.L. Hench, Bioactive material,. Ceramics International, **22** (1996) 493-507. https://doi.org/10.1016/0272-8842(95)00126-3

[10]   Lavernia, C. and J.M. Schoenung, CalciumPhosphate Ceramics as Bone Substitutes, Ceram.Bull,. **70** (1991) 95-100

[11]   Sergo, V., Sbaizero, O. and Clarke, D.R., Mechanical and chemical consequences of the residualstresses in plasma sprayed hydroxyapatite coatings, Biomaterials, **18** (1997) 477-482. https://doi.org/10.1016/S0142-9612(96)00147-0

[12]   Constantz, B.R., et al., Skeletal repair by in situ formation of the mineral phase of bon,. Science, **24** (1995) 1796-1799. https://doi.org/10.1126/science.7892603

[13]   Hench, L.L. and J. Wilson, in An Introduction to Bioceramics, L.L. Hench and J. Wilson, Editors., Singapore: World Scientific Publishing Co. Pte. Ltd, (1993). https://doi.org/10.1142/2028

[14]   Vogel, W. and W. Holand, The development of Bioglass Ceramics for medical applications, Angewandte Chemie International Edition, **26** (1987) 527-544. https://doi.org/10.1002/anie.198705271

[15]   Pajamäki, K.J.J., et al., Bioactiveglass and glass-ceramic-coated hip endoprosthesis: experimental study in rabbit, J. Mater. Sci.: Mater. Medi., **6** (1995) 14-18. https://doi.org/10.1007/BF00121240

[16]   Ohtsuki, C., T. Kokubo, and T. Yamamuro, Mechanism of apatite formation on $CaO-SiO_2.P_2O_5$glasses in a simulated body fluid.Journal of Non-Crystalline Solids, **143** (1992) 84-92. https://doi.org/10.1016/S0022-3093(05)80556-3

[17]   Posner, A.S., in Technology, Biological Functions, and Applications, J.R.V. Wazer, Editor., Interscience Publishers, Inc., New York, (1961)

[18]   Mellor, J.M., Comprehensive Treatise on Inorganic and Theorical Chemistry, London: Longmans Green, (1922)

[19]   Narasaraju, T.S.B. and D.E. Phebe, Some physico-chemical aspects of hydroxyapatite, J. Mater. Sci., **31** (1996) 1-21. https://doi.org/10.1007/BF00355120

[20]   Mohammadi, S., et al., Cast titanium as implant material. Journal of Materials Science: Materials in Medicine, **6** (1995) 435-444. https://doi.org/10.1007/BF00123367

[21]   Le Geros, R.Z. and J.P. LeGeros, in An Introduction Bioceramics, L.L. Hench and J. Wilson, Editors, Singapore: World Scientific Publishing Co. Pte. Ltd., (1993)

[22]   Aza, P.N.D., et al., Bioceramics- somulated body fluid interfaces: pH and its influence of hydroxyapatite formation, Journal of Materials Science: Materials in Medicine, **7** (1996) 399-402. https://doi.org/10.1007/BF00122007

[23]   Galliano, P.G. and J.M.P. Lopez, Thermal behaviour of bioactive alkaline-earth silicophosphate glasses. Materials Science:materials in medicine, **6** (1995) 353-359. https://doi.org/10.1007/BF00120304

[24]   Holmes, R.E. and S.M. Roser, Poroushydroxyapatite as a bone graft substitute in alveolar ridge augmentation: ahistometric study, Int. Jour. Oral Maxilifac. Surg., **16** (1987) 718-728. https://doi.org/10.1016/S0901-5027(87)80059-0

[25]   Lange, G.L.D., et al., A clinical, radiographic, and histological evaluation of permucosal dental implants of dense hydroxylapatite in dogs, J. DentRes., **68** (1989) 509-518. https://doi.org/10.1177/00220345890680031601

[26]   Page, D.G. and D.M. Laskin, Tissue responseat the bone-implant interface in a hydroxylapatite augmented mandibular ridge, Jour. Oral Maxillofac Surg., **45** (1987) 356-358. https://doi.org/10.1016/0278-2391(87)90360-0

[27]   Hench, L.L., Bioceramics: From Concept to Clini,. Journal of the American Ceramic Society, **74** (1991) 1487-1510. https://doi.org/10.1111/j.1151-2916.1991.tb07132.x

[28]   Frame, J.W., P.G.J. Rout, and R.M. Browne, Human tissue response to porous hydroxyapatite implants, A case report. Int.Jour. Oral Maxilifac. Surg., **18** (1989) 142-144. https://doi.org/10.1016/S0901-5027(89)80111-0

[29]   Stea, S., et al., Quantitative analysis of the bone-hydroxypatite coating interface, Journal of Materials Science: Materials in Medicine, **6** (1995) 455-459. https://doi.org/10.1007/BF00123370

[30]   Wang, P.E. and T.K. Chaki, Hydroxyapatite films on silicon single crystals by a solution technique: texture, supersaturation and pH influence, Journal of Materials Science: Materials in Medicine, **6** (1995) 94-104. https://doi.org/10.1007/BF00120415

[31]   Ducheyne P , S. Radin, M.Heughebaert, J.C.Heughebaert, Calcium phosphate ceramic coating on porous titanium: Effect of structure and composition on electrophoretic deposition, vacuum sintering and in vitro dissolution, Biomaterials, **19** (1990) 244-254. https://doi.org/10.1016/0142-9612(90)90005-B

[32]   Ducheyne, P., et al., The effect of hydroxyapatite impregnation on skeletal bonding of porous coated implants. J. Biomed. Mater Res., **14** (1980) 225-237. https://doi.org/10.1002/jbm.820140305

[33]   Ducheyne, P. and J.M. Cuckler, Bioactive ceramic prosthetic coating, Clin. Orthop. Relat. Res., **276** (1992) 102-14. https://doi.org/10.1097/00003086-199203000-00014

[34]   Ducheyne, P. and K.E. Healy, The effect of plasma-sprayed calcium on the metal ion release from porous chromium alloys, Journal of Biomedical Materials Research, **22** (1988) 1137-1163. https://doi.org/10.1002/jbm.820221207

[35]   Ogino, M. and L.L. Hench, Formation of calcium phosphate films on silicate glasses, J. Non-Cryst. Solids, **38 & 39** (1980) 673-678. https://doi.org/10.1016/0022-3093(80)90514-1

[36]   Hill, R., Apatite-mullite glass-ceramics. J Mater Sci, **6** (1995) 311-318. https://doi.org/10.1007/BF00120298

[37]   Yaszemski, M.J., et al., Evolution of bone transplantation: molecular, cellular and tissue strategies to engineer human bone, Biomaterials, **17** (1996) 175-185. https://doi.org/10.1016/0142-9612(96)85762-0

[38]   Liu, H.S., et al., Hydroxyapatite synthesized by a simplified hydrothermal method. Ceram. Int., **23** (1997) 19-25. https://doi.org/10.1016/0272-8842(95)00135-2

[39]   Fernades, G.F. and M.C.M. Laranjeira, Calcium Phosphate Biomaterials from Marine Alga, Hydrothermal Synthesis and Characterisation, Química Nova, **23** (2000) 441-446. https://doi.org/10.1590/S0100-40422000000400002

[40]   Bet, M.R., G. Goissis, and A.M.D.G. Plepis, Compósitos Colágeno Aniônico: Fosfato de Cálcio, Preparação e Caracterização. Química Nova, **20** (1997) 475-477. https://doi.org/10.1590/S0100-40421997000500006

[41]   Heimke, G., Advanced ceramics for biomedical application.Angewandte Chemie, **101** (1989) 111-116. https://doi.org/10.1002/ange.19891010141

[42]   Rhee, S.H. and J. Tanaka, Hydroxyapatite coating on a collagen membrane by a biomimetic method, J. Am. Ceram. Soc., **81** (1998) 3029-3031. https://doi.org/10.1111/j.1151-2916.1998.tb02734.x

[43]   Doi, Y., et al., Formation of apatite-collagen complexes, J. Biomed. Mater. Res., **31** (1996) 43-49. https://doi.org/10.1002/(SICI)1097-4636(199605)31:1<43::AID-JBM6>3.0.CO;2-Q

[44]   Bigi, A., S. Panzavolta, and N. Roveri, Hydroxyapatite-gelatin films: a structural and mechanical characterization. Biomaterials, **19** (1998) 739-744. https://doi.org/10.1016/S0142-9612(97)00194-4

[45]   Liu, Q., J.R.D. Winjn, and C.A.V. Blitterswijk, Covalent bonding of PMMA, PBMA and poly(HEMA) to hydroxyapatite particles, J. Biomed. Mater. Res., **40** (1998) 257-263. https://doi.org/10.1002/(SICI)1097-4636(199805)40:2<257::AID-JBM10>3.0.CO;2-J

[46]   Nimni, M.E. and R.D. Harkness, Collagen: Biochemistry, biomechanics, biotechnology, United States: CRC Press, 1 (1988)

[47]   Nimni, M.E., Collagen: Biochemistry, biomechanics, biotechnology, United States: CRC Press, 3 (1988)

[48]   Fukada, E., in Ferroelectric Plymers: Chemistry, Physics and Application, H.S. Nalwa, Editor, Marcel Dekker, Inc.: New York, (1995)

[49]   Mascarenhas, S., in Topics in Applied Physics, Springer-Verlag: Berlin, (1987)

[50]   Catauroa, M., A. Dell'Erab, and S.V. Cipriotic, Synthesis, structural, spectroscopic and thermoanalytical study of sol–gel derived SiO_2–CaO–P_2O_5 gel and ceramic materials, Thermochimica Acta, **625** (2016) 20-27. https://doi.org/10.1016/j.tca.2015.12.004

[51]   Nandi, S.K., B. Kundu, and S. Datta, Development and Applications of Varieties of Bioactive Glass Compositions in Dental Surgery, Third Generation Tissue Engineering, Orthopaedic Surgery and as Drug Delivery System, Biomaterials Applications for Nanomedicine, **4** (2011) 69-116

[52]   Jones, J.R., et al., Bioglass and Bioactive Glasses and Their Impact on Healthcare, International Journal of Applied Glass Science, **7** (2016) 423-434. https://doi.org/10.1111/ijag.12252

[53]   Kaur, G., et al., A review of bioactive glasses: Their structure, properties, fabrication, and apatite formatio, Journal of Biomedical Materials Research Part A, **102** (2014) 254-274. https://doi.org/10.1002/jbm.a.34690

[54]   Krishnan, V. and T. Lakshmi, Bioglass: A novel biocompatible innovation, J Adv Pharm Technol Res., **4** (2013) 78-83. https://doi.org/10.4103/2231-4040.111523

[55]   Hench, L., The story of Bioglass, J Mater Sci: Mater Med, **17** (2006) 967-978. https://doi.org/10.1007/s10856-006-0432-z

[56]   Crovace, M.C., et al., Biosilicate - A multipurpose, highly bioactive glass-ceramic. In vitro, in vivo and clinical trialsBiosilicate - A multipurpose, highly bioactive glass-ceramic,In vitro, in vivo and clinical trials,Journal of Non-Crystalline Solids, **432** (2016) 90-110. https://doi.org/10.1016/j.jnoncrysol.2015.03.022

[57]   Bramhill, J., S. Ross, and G. Ross, Bioactive Nanocomposites for Tissue Repair and Regeneration: A Revie, Int. J. Environ. Res. Public Health, **14** (2017) 1-21. https://doi.org/10.3390/ijerph14010066

Materials Research Foundations **57** (2019) 1-22   doi: https://doi.org/10.21741/9781644900390-1

[58] Peitl, O., et al., Compositional and microstructural design of highly bioactive P_2O_5-Na_2O-CaO-SiO_2 glass-ceramic,Acta Biomaterialia, **8** (2012) 321-332. https://doi.org/10.1016/j.actbio.2011.10.014

[59] Martınez, A., I. Izquierdo-Barba, and M. Vallet-Regi, Bioactivity of a CaO-SiO_2 Binary Glasses System, Chemistry Materials, **12** (2000) 3080-3088. https://doi.org/10.1021/cm001107o

[60] Jones, J.R., Review of bioactive glass: From Hench to hybrids, Acta Biomaterialia, **9**(2013) 4457-4486. https://doi.org/10.1016/j.actbio.2012.08.023

[61] Rahaman, M.N., et al., Bioactive glass in tissue engineering, Acta Biomater., **7** (2011) 2355–2373. https://doi.org/10.1016/j.actbio.2011.03.016

[62] Hongxin, W., et al., Preparation and Characterization of the System SiO2-CaOP2O5 Bioactive Glasses by Microemulsion Approach, Journal of Wuhan University of Technology-Mater. Sci. Ed., **28** (2013) 1053-1057. https://doi.org/10.1007/s11595-013-0818-y

[63] Lee, J.-H., S.-J. Seo, and H.-W. Kim, Bioactive glass-based nanocomposites for personalized dental tissue regeneration, Dent Mater J, **35**(5) (2016) 710-720. https://doi.org/10.4012/dmj.2015-428

[64] Mabrouk, M., et al., Comparative Study of Nanobioactive Glass Quaternary System 46S6, Bioceramics Development and Applications, **4** (2014) 1-4

[65] Mezahi, F.-Z., et al., Reactivity kinetics of 52S4 glass in the quaternary system SiO_2–CaO–Na_2O–P_2O_5: Influence of the synthesis process: Melting versus sol–gel, Journal of Non-Crystalline Solids, **361** (2013) 111-118. https://doi.org/10.1016/j.jnoncrysol.2012.10.013

[66] Gestel, N.A.P.V., et al., Clinical Applications of S53P4 Bioactive Glass in Bone Healing and Osteomyelitic Treatment: A LiteratureReview, BioMed Research International, (**2015**) 1-12. https://doi.org/10.1155/2015/684826

[67] Forsback, A.-P., S. Areva, and J.I. Salonen, Mineralization of dentin induced by treatment with bioactive glass S53P4 in vitro, Acta Odontol Scand, **62** (2004) 14-20. https://doi.org/10.1080/00016350310008012

[68] Bizari, D., et al., Synthesis, characterization and biological evaluation of sol-gel derived nanomaterial in the ternary system 64 % SiO_2 - 31 % CaO - 5 % P_2O_5as a bioactive glass: in vitro, Ceramics – Silikáty, **67** (2013) 201-209

[69] Liu, J. and X. Miao, Sol–gel derived bioglass as a coating material for porous aluminascaffolds, Ceramics International, **30** (2004) 1781-1785. https://doi.org/10.1016/j.ceramint.2003.12.120

Eng. Magnetic, Dielectric and Microwave Properties of Ceramics and Alloys   Materials Research Forum LLC
Materials Research Foundations **57** (2019) 1-22   doi: https://doi.org/10.21741/9781644900390-1

[70]   Ma, J., et al., Influence of the sintering temperature on the structural feature and bioactivity of sol–gel derived SiO_2–CaO–P_2O_5 bioglass, Ceramics International, **36** (2010) 1911–1916. https://doi.org/10.1016/j.ceramint.2010.03.017

[71]   Mortazavi, V., et al., Antibacterial effects of sol-gel-derived bioactive glass nanoparticle onaerobic bacteria, Journal of Biomedical Materials Research Part A, **94** (2010) 160-168. https://doi.org/10.1002/jbm.a.32678

[72]   Saravanapavan, P., et al., Bioactivity of gel– glass powders in the CaO-SiO_2 system: A comparison with ternary (CaO-P_2O_5-SiO_2) andquaternary glasses (SiO_2-CaO-P_2O_5-Na_2O), Journal of Biomedical Materials Research Part A, **66** (2003) 110-119. https://doi.org/10.1002/jbm.a.10532

Eng. Magnetic, Dielectric and Microwave Properties of Ceramics and Alloys   Materials Research Forum LLC
Materials Research Foundations **57** (2019) 23-56   doi: https://doi.org/10.21741/9781644900390-2

Chapter 2

Lead Hexaferrite - A Brief Review

S.A. Palomares-Sánchez[1], M.I. González Castro[2], S. Ponce Castañeda[3]

[1]Facultad de Ciencias, UASLP (FC-UASLP). 78000 San Luis Potosí, México

[2]Universidad Autónoma de Chihuahua (UACh). 31110 Chihuahua, Chi. México

[3]Universidad Politécnica de San Luis Potosí (UPSLP), 78369 San Luis Potosí, México

sapasa04@fciencias.uaslp.mx

Abstract

Lead hexaferrite belongs to the family of compounds with the chemical formula $MFe_{12}O_{19}$ (M = Ca, Ba, Sr, Pb, La). One of the main reasons the lead hexaferrite has not been exhaustively studied is because its magnetic properties are inferior to barium and strontium hexaferrites. Few studies have been carried out after the description of its structure, in 1938; nevertheless, one of its advantages is that it can be prepared at lower temperatures than conventional reported ferrites and it is worth studying its properties when compared with other members of the family. Therefore, this work deals with a brief review of preparation methods, properties and applications of this compound.

Keywords

Lead Ferrites, Crystal Structure, Preparation Methods, Magnetic Properties

Contents

Eng. Magnetic, Dielectric and Microwave Properties of Ceramics and Alloys   Materials Research Forum LLC
Materials Research Foundations **57** (2019) 23-56   doi: https://doi.org/10.21741/9781644900390-2

1.   Introduction

A large group of oxide ceramic magnets consists of hexagonal ferrites, or hexaferrites. These compounds have a hexagonal and rhombohedral symmetry and are classified as M, Y, W, X, U and Z type. M-type hexaferrites form a family of magnetic materials with formula $MFe_{12}O_{19}$ (M = Ca, Ba, Sr, Pb, La) [1]. The production of this kind of materials, used as permanent magnets, exceeds that of any magnetic material. Among the M type, $BaFe_{12}O_{19}$ (BaM) and $SrFe_{12}O_{19}$ (SrM), are of the most important materials due to its numerous applications as permanent magnets in loudspeakers, motors, microwave devices, recording media, etc. [2]. The most important magnetic oxide in this family is the barium hexaferrite, $BaFe_{12}O_{19}$ [3]. One non-exhaustive nor systematic search in Academic Google of "lead hexaferrite", "strontium hexaferrite" and "barium hexaferrite" yielded only 121, 2130 and 4870 result, respectively, indicating the scarcity of works related with PbM. On the other hand, lead hexaferrite ($PbFe_{12}O_{19}$) has recently received more attention due to its possible multiferroic properties at room temperature, as a single phase compound, for applications in magnetoelectronics [4] [5]. One of the main reasons PbM has not been exhaustively studied is because lead is considered harmful for the environment [6] and sometimes it has not been possible to obtain the pure phase.

Recently, very good reviews and books on the general aspects of hexagonal ferrites have been published [1], [7], [8], but they are not focused in PbM. In this work, the description of the main properties of PbM, followed by the preparation methods, as powder and film, are presented in chronological order. Finally, substituted and composites of PbM are listed.

2.   Structure

The structure of the $PbFe_{12}O_{19}$ is the prototype of the M-type hexaferrites. It is isostructural with magnetoplumbite, whose natural form has the composition $PbFe_{7.5}Mn_{3.5}Al_{0.5}Ti_{0.5}O_{19}$ [1]. The first report on the structure of the PbM was made in 1938 [9]. The structure can be considered as built of R and S blocks, with 10 oxygen layers in the unit cell. Hexagonal R block is built by three oxygen layers with one oxygen ion replaced by Pb^{2+} and with the composition $(Ba^{2+}Fe_6^{3+}O_{11}^{2-})^{2-}$, whereas the S block, with two spinel units, has the composition $(Fe_6^{3+}O_8^{2-})^{2+}$. Then, the unit cell of PbM can be expressed as RSR*S*, where * indicates a rotation of 180° of R and S around the c-axis [10]. The crystallographic parameters of the PbM are listed in Table I [11].

The unit cell is formed by two stacked formula units, $2(PbFe_{12}O_{19})$, one with S and R blocks and the other one with S* and R*, Figure 1. Each unit has 32 atoms, *i. e.*, 24 Fe^{3+} ions, 38 O^{2-} ions and 2 Pb^{2+} ions. In one formula unit, there are 12 Fe^{3+} ions, out of them, 9 occupy the octahedral sites, two occupy tetrahedral sites and one is a 5-fold coordinated. Fe^{3+} ions are located in five types of sublattices as indicated in Table II. There is a total of 20 μB per formula unit at 0 K [12].

Table I. Unit cell properties of PbM [11].

Formula	$PbFe_{12}O_{19}$
Space group	$P6_3/mmc$
Z	2
a (Å)	5.873(2)
c (Å)	23.007(6)
V (Å^3)	687.2(1)
ρ_{cal} (g/cm^3)	5.71

Table II. Fe sublattices of PbM.

Sublattice	Number of sites	Symmetry	Block	Spin
k	12	octahedral	Shared R and S	$(3\downarrow)(3\downarrow)$
a	2	octahedral	S	$1\uparrow$
b	2	bipyramidal	R	$1\uparrow$
f_2	4	octahedral	R	$2\downarrow$
f_1	4	tetrahedral	S	$2\downarrow$

3.   Preparation methods

Several preparation procedures have been developed or used to obtain hexaferrites. Usually, the techniques to prepare BaM and SrM have been later used to prepare PbM. Among them, are the ceramic method [13], chemical coprecipitation [14], sol-gel technique [15], mist pyrolysis method [16], reverse emulsion technique [17], aerosol pyrolysis [18], Pechini method [19], glass crystallization [20], the flux growth method [11], self-propagating high-temperature synthesis [21], citrate precursor method [22], etc. The evolution of the research of the PbM indicates that at the beginning the main objective was to study the complete system Fe-Pb-O. Thereafter, the preparation method used to obtain SrM and BaM was also used to obtain PbM as single crystals, powders, thin films, nanostructures and composites. In some works, it is reported the occurrence of secondary phases, like iron oxides, due to the loss of lead during thermal treatments. Then, several attempts were made to obtain the pure phase of the PbM by using different

Eng. Magnetic, Dielectric and Microwave Properties of Ceramics and Alloys   Materials Research Forum LLC
Materials Research Foundations **57** (2019) 23-56   doi: https://doi.org/10.21741/9781644900390-2

ratios of Pb/Fe and heat treatment temperatures. Also, its structural, magnetic, electrical, optical etc. properties were studied along.

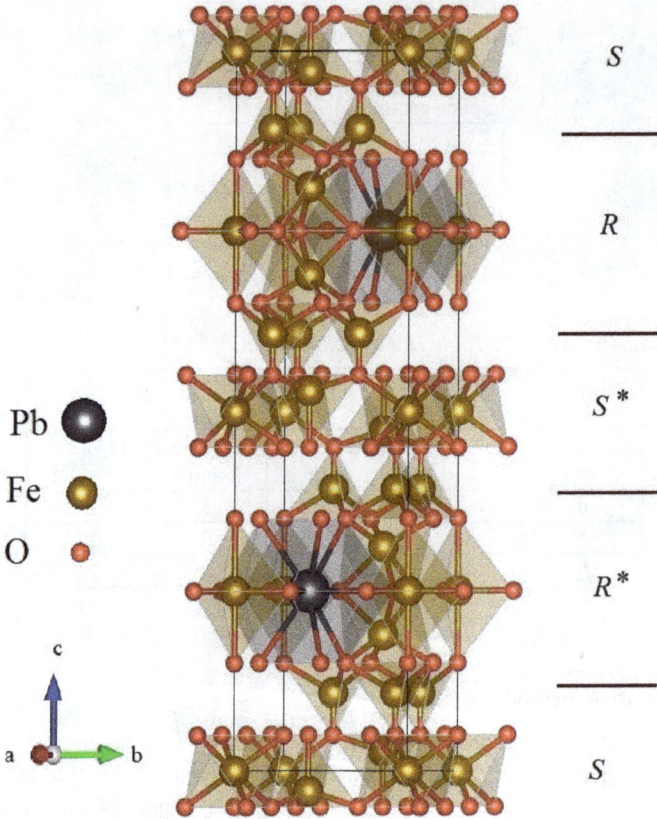

Figure 1 Unit cell of PbM.

Eng. Magnetic, Dielectric and Microwave Properties of Ceramics and Alloys   Materials Research Forum LLC
Materials Research Foundations **57** (2019) 23-56   doi: https://doi.org/10.21741/9781644900390-2

Table III. Atomic fractional coordinates in PbM [9].

Atom	Sublattice	x	Y	z
Pb1	2d	0.3333	0.6666	0.7500
Fe1	2a	0.0000	0.0000	0.0000
Fe2	2b	0.0000	0.0000	0.2500
Fe3	$4f_1$	0.3333	0.6666	0.0278
Fe4	$4f_2$	0.3333	0.6666	0.1889
Fe5	12k	0.1667	0.3333	0.1083
O1	4e	0.0000	0.0000	0.0015
O2	4f	0.3333	0.6666	-0.0500
O3	6h	0.1860	0.3720	0.2500
O4	12k	0.1666	0.3333	0.0500
O5	12k	0.5000	1.0000	0.1500

A recent review on the system Fe-Pb-O has been published [23], however, the first known report on the obtaining of PbM by heating the powders coprecipitated from solutions of nitrates was reported by Adelsköld, as quoted by Kojima [10]. After that, several studies have been carried out on the compounds formed in the system PbO-Fe_2O_3 [24] [25][26] [27].

In [24], cylindrical samples of PbM were prepared by mixing PbO and α-Fe_2O_3 with different ratios and pressed at 294.2×10^6 Pa. Thereafter, the samples were heated in order to determine the phase diagram. PbM is formed by according to the reaction 7(PbO·Fe_2O_3)$\rightarrow$ 2(PbO·6Fe_2O_3) + 2(2PbO·Fe_2O_3) + PbO at temperatures above 825 °C. On the other side, when the stoichiometric mixture PbO:6Fe_2O_3 was heated at 850 °C the MPb was formed. However, it was also determined that at 950 °C there was a loss of weight of the sample do to evaporation of lead. These results were confirmed in [25], but the PbM was not found in [27] in the same system.

After these seminal works, where the formation of the PbM was confirmed, other authors prepared directly the compound and several studies on the magnetic, optical, electric, etc. were also carried out.

Tokar [28] prepared a series of samples ranging from PbO·4Fe_2O_3 to PbO·6.5Fe_2O_3 using high purity Fe_2O_3 and PbO by the usual ceramic method. Cylindrical samples were pre-fired at different temperatures (850 °C–1225 °C). After grinding the disks and pressed again as cylindrical samples, they were sintered at temperatures from 850 °C to 1300 °C. The sintering times varied from 15 min to 120 min. Likewise, by adding silica to the PbO·5Fe_2O_3 specimen, it was observed two phases; *i.e.* Fe_2O_3 and PbO·6Fe_2O_3, separated

Eng. Magnetic, Dielectric and Microwave Properties of Ceramics and Alloys   Materials Research Forum LLC
Materials Research Foundations **57** (2019) 23-56   doi: https://doi.org/10.21741/9781644900390-2

by a grain boundary phase of lead silicate. Adding silica and boria raised the energy product to 1.4×10^6 G·Oe.

Single crystals of BaM, SrM and PbM were grown by the flux method. PbO was used for the preparation of PbM. The Faraday effect and optical absorption were measured on polished (001) surfaces while the Kerr effect and reflection were measured on unpolished (001) surfaces.

Ram et al. [30] studied the crystallization process of the magnetic phases in the system $50PbO-20Fe_2O_3-30B_2O_3$. The samples were prepared by melting and casting at 1,400 K and then heat-treated in two successive steps at different combinations of nucleation and growth temperatures. In all the samples, the main crystalline phase was PbM. The maximum coercivity was obtained in the compound $Pb_5Fe_{14}O_{26}$ with a value of ~3000 Oe and, for PbM, of ~2100 Oe.In another kind of study, magneto-optical Kerr effect was determined at 290 K, between 2.0 eV and 5.5 eV, and the calculation off-diagonal permittivity tensor elements was carried out on PbM [31].

Also, the flux growth method was used to grow crystals of PbM. The initial temperature was 1000 °C and the crystal were grown by slow cooling [11]with the objective of studying the disordering of Pb as a possible consequence of a lone pair-bond pair interaction. In [32] the crystals were used to obtain ^{57}Fe NMR spectra of single crystals of PbM at 4.2 K. It was found that Al enters the 2a site preferably.

Carp et al. reported the preparation of PbM using nonconventional methods were $Pb(NO_3)_2$ and $FeC_2O_4 \cdot 2H_2O$ were used as raw materials [33]. The goal of this study was determined by the mechanism of PbM formation. The precursors were mixed in a ratio $FeC_2O_4 \cdot 2H_2O/Pb(NO_3)_2 = 12$ in hot water and NaOH was added to obtain a pH~ 13. After stirring, the precipitate is filtered and washed with water. The precursors were prepared in three different processes. After that, thermal treatments were carried out at 600 °C, 800 °C and 1000 °C. The significant results in this work are that all the samples contain small amounts of α-$Fe_2O_3$and that 800 °C is the optimum temperature to obtain the PbM.

The fluxed-solution method was used to grow single crystals in order to study the magnetic relaxation of oscillating domain walls in PbM [34].

The mist pyrolysis method was also used to prepare PbM [35]. An aqueous solution of distilled water, $Pb(NO_3)_2$ and $FeC_2O_4 \cdot 9H_2O$ was used as starting material. The mist was produced by an ultrasonic nebulizer vibrating at 2.6 MHz. Then, it was carried by a flux of oxygen across a furnace at 800 °C. PbM was obtained with a molar ratio Pb/Fe in the range 0.13~0.23. Using the stoichiometric ratio (0.083), a mixture of PbM and Fe_2O_3 is obtained. The magnetic properties are $\sigma_s = 56$ emu/g, $_iH_c = 4\sim5$ kOe, and $(BH)_{max} = 0.7$-0.9 MG-Oe.

Díaz-Castañón et al. prepared the PbM by two methods; i.e. chemical coprecipitation and metal-organic precursors (sol-gel) [36]. In the former method, coprecipitated salts were obtained by dripping a solution of $Pb(NO_3)_2$ and $Fe(NO_3)_3 \cdot 9H_2O$, with a stoichiometric ratio Pb/Fe = 12, in a solution of $NaOH/Na_2CO_3$ with a $pH = 10.5$. Finally, the precipitates were washed, dried and powdered. For obtaining the metal organic powders, a solution of $Pb(NO_3)_2$ and $Fe(NO_3)_3 \cdot 9H_2O$ (Pb/Fe = 12) in ethylene glycol was heated and stirred until the organic compound was removed. Thereafter, both kinds of powders were heat treated at 920 °C for 2 hours. By using these procedures, no secondary phases were detected outer the PbM. The coprecipitated powders had $\sigma_s = 56$ emu/g, $_iH_c = 1.6$ kOe, and $H_a = 13.4$kOe, whereas the MOD powders had $\sigma_s = 54$ emu/g and $_iH_c = 5.0$ kOe. The temperature was the same for the two kinds of samples, 721 K.

The ceramic method was also used to prepare the PbM for studies of diffusion transport in hexagonal ferrites [37]. For the preparation of the single phase PbM grade PbO and Fe_3O_2 were used as precursors. According to X-Ray analysis, the single phase of PbM was obtained. A high Pb mobility along structurally disordered layers was observed by means of the radio-tracer method with ^{212}Pb as tracer ($\tau_{1/2} = 10.64$ h, $\beta = 0.355$ MeV, 0.589 MeV, $\gamma = 0.239$ MeV, 0.300 MeV).

Nanosized PbM was prepared by the citrate precursor method [38]. Two solutions were prepared: lead citrate and a ferric citrate solution. The former one was prepared from stoichiometric amounts of lead nitrate and citric acid and mixed with the later in proportion for obtaining the PbM. The citrate precursor was obtained from the solution, with a $pH = 2.6$, after it was refluxed at 100 °C for 12 hours and later dehydrated from ethanol. The precursor was then decomposed in the temperature range from 500 °C to 800 °C. The nanoparticles of PbM were in the size range from 6.0 nm to 25.0 nm. The maximum value of H_{ic} is 0.471 T for particle size of 17.2 nm, close to the theoretical limit for isotropic nanocrystalline lead hexaferrite.

The self-propagating high-temperature synthesis was used to prepare BaM, SrM and PbM to study the influence of both the combustion and sintering temperatures on the defect density and internal stress in these ferrites [39]. For the PbM, the precursors were lead oxide and iron oxide. The powders were mixed and sintered in a quartz reactor and afterward ground in a vibratory mill at a sample/ball ratio of 1:10. After adding 0.5 % triethanolamine, samples were dried and polyvinyl alcohol was added. Cylindrical samples were pressed at 100 MPa and subsequently heated at 1000-1200 °C. The measured properties of PbM are remanent induction, $B_r = 0.19$ T, magnetization

Eng. Magnetic, Dielectric and Microwave Properties of Ceramics and Alloys   Materials Research Forum LLC
Materials Research Foundations **57** (2019) 23-56   doi: https://doi.org/10.21741/9781644900390-2

coercivity, H_M = 190 kA/m, induction coercivity, H_B = 132 kA/m, energy product, $(BH)_{max}$ = 0.19 T, conductivity, ρ = $4{\times}10^7 \Omega{\cdot}m$, hardness, HRA = 74, compressive strength, σ_c = 190 MPa, tensile strength, σ_t = 21 MPa, bending strength, σ_b = 58 MPa.

PbM was also prepared by using modifications to the traditional ceramic route [40]. The precursors were high purity PbO and Fe_2O_3. For the pre-sintering treatment, six temperatures, ranging from 850 °C to 1150 °C, were chosen. The pure phase of PbM corresponds to the sample pre-sintered at 900 °C and sintered at 1200 °C. For MPb presintered at 900 °C and sintered at 1200 °C/2 h, M_s = 62.0 emu/g and $_iH_c$ = 4.0 kOe. In this work, the modification is the reduction of the pre-sintering temperature to 900 °C.

It was demonstrated that the principal losses of PbO occurred during the pre-sintered treatment. In order to calculate the losses of lead during thermal treatments, the Rietveld refinement method was used as an intermediate step in the ceramic method [13]. The samples were prepared by the traditional ceramic method. After the pre-sintering stage, the Rietveld refinement was used in order to quantify the phases present (PbM and Fe_2O_3) to quantify the losses of Pb; that is, the quantity of unreacted hematite. It was found that it is necessary to add 1.5 %, in weight, in excess, of lead oxide on the stoichiometry to obtain the pure phase of PbM, Figure 2. The magnetic properties of PbM were M_s = 60.0 A·m^2/kg and $_iH_c$ = $3.06{\times}10^5$ A/m.

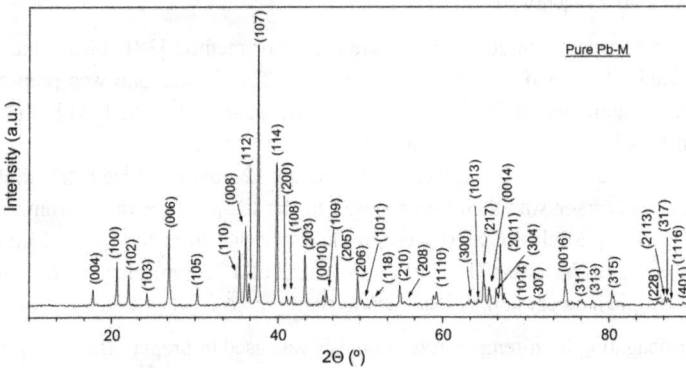

Figure 2 Diffractogram of PbM obtained with Co Kα radiation (1.788996 Å) [13].

A sol-gel route was also used to prepare thin films and bulk PbM [15]. 2-butoxyethanol was used as a solvent of the solution of iron $(Fe(NO_3)_3{\cdot}9H_2O)$ and lead $(Pb(CH_3CO_2)_2{\cdot}3H_2O)$. Two molar ratios were used (Fe/Pb = 12:1 and 12:1.5). The resulting precipitate (lead nitrate) was redissolved with acetylacetone. The solution was heated at 350 °C and the thermal treatments of the resulting dry powders were between 500 °C and 1000 °C. With this method, it was possible to form the PbM already at 650 °C. The film of PbM deposited on monocrystalline $MgAl_2O_4$ (111) showed a magnetic moment ~15 μemu and H_c ~2500 Oe.

Also, the preparation of nanosized PbM is reported by using the citrate precursor method with some variations [41]. The pure PbM was obtained when the ratio Pb/Fe was 1:6. This ratio was chosen to synthesize the PbM at different temperatures, from 600 °C to 900 °C during 2 hours. In this work, it is reported that the crystalline PbM appeared already at 700 °C. A coercivity of 3.8 kOe and saturation magnetization of 49.0 emu/g are obtained from the sample calcined at 800 °C.

The citrate-nitrate gel combustion method was used to prepare PbM to study its structure and thermochemical behavior [42]. The samples were prepared by dissolving PbO and dried $Fe(NO_3)_3 \cdot 9H_2O$ in diluted nitric acid. The solution was heated in a hot plate at around 100 °C to remove water and oxides of nitrogen. The gel was heated at 177 °C to dryness. The powder was ground and heated at 827 °C during 100 hours with two intermediate grindings. The resulting powder corresponded to the pure phase of PbM. The Curie temperature for PbM was determined in 718 K. Average axial thermal expansion coefficients, from 298 K to 1273 K, and the melting temperature of PbM was $\alpha_a = 10.80 \times 10^6$ K^{-1}, $\alpha_c = 18.34 \times 10^6$ K^{-1}, $\alpha_v = 40.46 \times 10^6$ K^{-1} and 1538 K, respectively.

Singhal et al. [43] also reported the obtaining of PbM using the sol-gel method to investigate the shielding effect of KCl, KBr and KI on the morphology of hexaferrites. The coercivity and saturation magnetization for PbM were 2420 Oe and 46.44 emu/g, respectively whereas for the sample with KCl were 3780 Oe and 35.97 emu/g. The saturation magnetization was reduced to 0.06 for the sample with KI.

To study the multiferroic properties of PbM, Tan et al. prepared the PbM using the sol-gel method (polymer precursor method) [44]. The precursors were lead acetate and ferric acetylacetonate. In this work, the loss of Pb was considered by taking the ratio to lead iron lower than 1:12. The resulting powders were calcined at 800 °C and pressed as pellets to be sintered at 1000 °C. A remnant polarization and a coercive electric field of 33.5 μC/cm2 and 96 kV/m, respectively were found. The remnant magnetic polarization and the magnetic coercivity were 24.5 emu/g and 2168 Oe, respectively.

To simulate the stabilization of a lead-containing sludge, PbM was prepared through a thermal process using lead hydroxide and an iron sludge [45]. Also iron and lead oxides were used to prepare PbM to compare the differences between pure oxide and sludge samples in phase transformation and morphology during thermal treatments. The

$PbFe_{12}O_{19}$ was the major phase at 1050 °C and 1100 °C, in sludge and oxide samples, respectively.

Nanoplates and nanoparticles of PbM were prepared using maltose as reductant [46]. The preparation method was the sol-gel auto-combustion using iron nitrate, lead nitrate and maltose. Three different molar ratios Pb/Fe were used; *i. e.* 1:6, 1:9 and 1:12. The final powders were calcined at different temperatures, from 600 °C to 900 °C. The combination of the ratios Pb/Fe and Pb/ maltose and different sintering temperatures resulted in the formation of nanoplates (Pb/Fe = 1:6, Pb/maltose = 1:13, 900 °C) and nanoparticles (Pb/Fe = 1:6, Pb/maltose = 1:26, 600 °C) of PbM.For the sample with Pb/Fe = 1:6, Pb/maltose = 1:26, 900 °C values of M_r = 0.205 emu/g and H_c = 5609 Oe were obtained, however, in the sample without maltose, these values were M_r = 0.130 emu/g and H_c = 2122Oe.

The ceramic and chemical coprecipitation methods were used to prepare the pure phase of PbM [47]. In the ceramic method α-$Fe_2O_3$and Pb_3O_4 were used as precursors. By adding 33.09 % wt. of lead oxide, calculated by the Rietveld refinement method, to the initial mixture, it was possible to obtain the pure PbM. The sintering temperature was 95 °C. In the chemical coprecipitation method, the precursors were iron and lead nitrates. The precipitated powders were sintered at 700 °C. It was found that morphologic characteristics and the microstructure of both kinds of samples influence on the magnetic properties. The magnetic properties of the ceramic samples are χ_i = 0.0129, σ_s = 72.0 emu/g, σ_r = 39 emu/g and H_c = 1.7 kOe, and for the coprecipitated samples are χ_i = 0.0031, σ_s = 38.3 emu/g, σ_r = 20.0 emu/g and H_c = 3.5 kOe.

The ceramic method was used to prepare BaM, SrM and PbM [5] and their multiferroic properties are reported. Only BaM and SrM exhibited ferroelectricity but not PbM.

Nanoparticles of PbM were prepared by using the sol-gel auto-combustion method with lead and iron nitrates as precursors and several carboxylic acids (malonic, succinic, malonic and maleic) as fuel and reducing capping agents to limit the size of the nanoparticles [48]. The maleic acid was the best fuel and capping agent to obtain pure PbM by using a ratio Pb/maleic acid = 1:13 and calcination temperature of 900 °C for 2 hours. For this sample, the specific saturation magnetization and coercivity are 27 emu/g and 1900 Oe, respectively and M_r/M_s = 0.57.

In order to determine its ferroelectric properties, pure PbM was prepared [4]. The compound was obtained by the polymer precursor method. A molar ratio lead/iron 1:9 was used to compensate for the lead losses during the thermal treatments. The remnant polarization is estimated to be 104 $\mu C/cm^2$ and the coercive field is measured to be 15.2

kV/m. The remnant magnetic moment was 30.8 emu/g, whereas the magnetization was 54.2 emu/g at a maximum applied field of 1.0 T.

A variation of the sol-gel auto-combustion method, with cherry juice as fuel and capping agent, was used to prepare nanoparticles of PbM [49]. The anthocyanin pigments of the cherry juice act as a reducing agent in a chemical reaction. In this work, the ideal amount of cherry juice to obtain pure PbM was 20 ml and calcination temperature of 800 °C for 2 hours. The particle size was 15 nm as determined by SEM. For this sample, the specific saturation magnetization and coercivity are 7.6 emu/g and 300 Oe, respectively.

Lead-hexaferrites and magnetic cellulose acetate nanocomposites were prepared [50] to study the magnetization, coercivity and remanence. The compounds used to prepare the samples were lead acetate, iron nitrate, ammonia, sodium hydroxide and ethylene glycol. The particle size and morphology of PbM was controlled by the number of microwave pulses applied to the initial solution. The power of the microwave was 170 W, 510 W and 850 W. The precipitate was calcined at temperatures from 550 °C to 850 °C. Sample calcined at 550 °C had a coercivity and saturation magnetization of 1830 Oe and 26.7 emu/g, respectively, whereas the prepared at 850 °C had a coercivity of 3200 Oe and saturation magnetization of 33 emu/g. The cellulose acetate-PbFe compound had a coercivity of about 3250 Oe and saturation magnetization of 3.7 emu/g.

In order to study the structural, magnetic and dielectric properties of nanoparticles of PbM [51], the sol-gel method was used. The powders were calcined at 700 °C, 750 °C, 800 °C, 850 °C, 900 °C, and 1000 °C with 1 h, 1.5 h, 2 h, 2.5 h and 3 h. The best annealing temperature and time was 800 °C and 3 h, respectively, to obtain pure PbM. The magnetic parameters for this sample were saturation magnetization, M_s = 43.08 emu/g, remanence, M_r = 25.60 emu/g, coercivity, H_c = 3865Oe, crystalline anisotropy field, H_a = 10101Oe, demagnetization field, H_d = 2049 Oe and magnetocrystalline anisotropy constant, K = 217.6 emu·kOe/g. Also, the results of dielectric properties of PbM are presented.

The sol-gel auto-combustion route was used to prepare $PbTiO_3/PbFe_{12}O_{19}$ nanocomposite. [52]. The nanocomposites presented superparamagnetism with a saturation magnetization of 3.5 emu/g.

The green hydrothermal method, with lemon extract as a surfactant, was used to prepare nano-spheres of PbM to be subsequently mixed with PbS to form a photo-catalyst [53]. The composites presented superparamagnetic behavior. The PbM had a saturation magnetization of 39 emu/g and a coercivity about 1300 Oe.

$PbFe_{12}O_{19}$ nanocomposites are successfully synthesized by auto-combustion sol-gel route by employing valine as reducing agent [54]. Three kinds of samples were prepared: PbM,

PbM/graphene and PbM/CNT. The results indicated that the optical and electrochemical properties of lead hexaferrites are improved in the nanocomposites. The magnetic properties of PbM are increased in the presence of graphene with $M_r = 21.0$ emu/g, $M_s = 34.5$ emu/g and $H_c = 5698$ Oe. Also, for the first time, the photocatalytic activity of lead hexaferrites is evaluated using the degradation of methyl orange under ultraviolet light irradiation.

The PbM was also prepared by sol-gel autocombustion and processed by microwaves [55] in order to study the multiferroic properties of $PbFe_{12-x}O_{19-\delta}$ ($x = 0.0, 0.25, 0.50, 0.75, 1.0$). The compounds used to prepare the PbM were $Pb(NO_3)_2$, $Fe(NO_3)_3 \cdot 9H_2O$, $(C_6H_8O_7)$, ethanol and ammonia. The molar ratios of $Pb(NO_3)_2$ to $Fe(NO_3)_3 \cdot 9H_2O$ were 1:12.00, 1:11.75, 1:11.50, 1:11.25 and 11.00. The resultant powders were pre-sintered at 650 °C for 30 min. Pellets- and toroidal shaped samples were finally sintered in a microwave furnace with a frequency of 2.45 GHz and a power output of 2.20 kW at 950 °C. X-ray analysis indicated that the pure phase of PbM was obtained with a crystallite size of 45 nm. Optical and electric properties were studied in these samples. Magnetic parameters were also reported for all the samples.

Nanoparticles of M-type lead hexaferrite ($PbFe_{12}O_{19}$) have been prepared using coprecipitation method in order to be studied using Mössbauer spectroscopy [14]. The hyperfine parameters such as line width, Γ, isomer shift, δ, electric quadrupole splitting, Δ, and magnetic hyperfine field, B_{hf} are reported indicating the superparamagnetic nature of the particles.

To study the influence of structural defects in lead ferrites on lead leaching [56], lead oxide and hematite were mixed and heat-treated at temperatures between 600 °C and 1000 °C for three hours and quenched in air at room temperature. As expected, one of the compounds was PbM, that was relatively free of structural defects and was found to be the preferred stabilization product to reduce the environmental hazard posed by lead in lead sludge.

For the first time, nanostructured $PbFe_{12}O_{19}$ was synthesized by the sonochemical method [57]. The starting materials were lead acetate, iron acetate and several capping agents, like cetyltrimethylammonium bromide, sodium dodecyl sulfate and polyethylene glycol. PbM was obtained by this route without further thermal treatment. A molar ratio $Pb^{2+}/Fe^{3+} = 6$, sonication time of 30 min, power of ultrasound waves of 60 W and the use of polyethylene glycol as capping agent was the best combination to obtain the PbM. A thermal treatment at 900 °C was used to increase both crystallinity and particle size. The pure PbM has a saturation and remaining magnetization of 26.73 emu/g and 13.94 emu/g, respectively, with $M_r/M_s = 0.52$. The intrinsic coercivity was about 1139.4 Oe.

4.   Films

Few reports on the preparation of lead hexaferrite as thin films have been published. One of the first reports on the preparation of thin films of hexaferrite with Pb, deposited in the oxidized silicon wafer, is that of Morisako [58]. The objective of this work was to reduce the temperature of deposition by substituting the barium by lead. The preparation technique was the dc diode magnetron sputtering. The composition of the films was $Pb_{0.55}Ba_{0.63}Fe_{12}O_x$. The coercivities $H_\perp$ and $H_\parallel$ and saturation magnetization, M_s, are 0.7-1.0 kOe, 0.2kOe and 250-300 emu/cm^3, respectively.

Dorsey et al. [59] deposited thin films on single crystal sapphire substrates (0001) by the pulsed laser deposition technique (PLD) using a pure polycrystalline PbM target. The composition of the deposited films was $PbFe_{12.9}O_{22.9}$. The film has magnetically isotropic behavior in the film plane with M_r/M_s) and H_c values of 88 ±2.9Oeand 2500± 97 Oe, respectively, but they are anisotropic with respect to the film normal and an estimate of the planar anisotropy field, H_A , is 77.5 kOe. The value of $4\pi M_s$ of the films is 630 Gauss at room temperature.

Also, oriented PbM was grown on (0001) sapphire substrates by the pulsed laser deposition technique [60] with the same conditions of oxygen pressure and temperature as in [61]. M_r/M_s, the saturation magnetization and the coercive field in these films are 0.86, 165 emu/cm^3 and 2.5 kOe, respectively.

Díaz- Castañón et al. [61] [62] used the pulsed laser ablation technique (PLD) to growth polycrystalline $PbFe_{12}O_{19}$ on Si/SiO_2 substrates. The target of $PbFe_{12}O_{19}$ was prepared by the ceramic method. The substrate temperature was 650 °C, 700 °C and 750 °C. Best conditions to obtain the PbM were 3.0 mbar of oxygen and 700 °C, lower than the necessary to obtain bulk PbM. The magnetic properties were M_s = 280 emu/cm^3, and $_iH_c$ = 3.8 kOe. The results of this experiment were used to demonstrate the validity of a model to study the formation of thin films of ternary oxides by pulsed laser deposition [63].

The same method to prepare bulk PbM was used to deposit thin films on $MgAl_2O_4$ (111) by the spin coating method [15]. After every deposition of the solution on the substrate, the sample was pyrolyzed at 350 °C and finally grown at 700 °C. The film of PbM showed a magnetic moment ~15 μemu and H_c ~2500 Oe.

(Ba·Pb) hexaferrite films were grown on sapphire (00l) by rf magnetron sputtering. [64]. Due to the lower diffusion activation barrier of Pb cations than that of Ba, crystallization at low temperature, 530 °C, was achieved. The composition of as-deposited (Ba·Pb) hexaferrite films were $(Ba_{0.55}Pb_{0.41})Fe_{12}O_{19}$. The grain size was ~40nm in diameter, $M_{s\perp}$ = 337 emu/cm^3 and $_iH_{c\perp}$ = 1.60 kOe.

Using the alternating target laser ablation deposition (ATLAD), with PbO and Fe_2O_3 as targets and MgO <111> as substrate [65], oriented thin films of PbM were grown. With this technique, it is possible to compensate for the loss of Pb by adjusting the number and energy of laser shots from the respective target. An unusual phase of $PbFe_{12}O_{19}$, where Pb^{2+} and Fe^{3+} ions are not within the same crystal plane, is proposed. The in-plane and out-of-plane coercive fields in the as-deposited films were measured to be 0.20kOe and 1.50kOe, respectively. After annealing the films at 900 ºC, these values changed to 0.18 kOe and 0.28 kOe. The saturation magnetization of the films was 1.8 kG.

5.   Substituted PbM

Modifying the properties of the PbM for various applications is one of the objectives of the research work in this material. The objectives range from decreasing the temperature in the deposition of thin films of hexaferritas until the absorption of microwaves. According to some reports, PbM could be a single phase multiferroic, making it a suitable candidate for spintronic and magneto-electronic applications. Other potential application is in the absorption of microwaves. This could be possible by substitutions of rare earth in hexaferrites in order to control its dielectric and magnetic properties. A list of the substitution in PbM is shown in Table II.

Table II. Composite/substituted PbM.

Compound	Preparation method	Objetive	Ref.
$PbAl_{12-x}Fe_xO_{19}$ $0 \leq x \leq 12$	Ceramic.	Determination of the limits of the solid solution, Correlation of the unit cell dimensions against composition.	[66]
$PbFe_{12-x}Ga_xO_{19}$ $x = 0.0 - 10.38$	Flux method.	Preparation of single crystals with this composition.	[67]
$Pb_{1.12}Sm_{0.04}Fe_{11.74}{}^{3+}Fe_{0.05}{}^{2+}O_{18.84}$ $PbSm_{0.08}Fe_{11.80}{}^{3+}Fe_{0.06}{}^{2+}O_{18.9}$	Flux method.	Replacement of Pb^{2+} by Sm^{3+}. Reduction in electrical resistance was observed.	[68]

$PbFe_{12-x}Ru_xO_{19}$ $x = 0.0, 0.02, 0.04, 0.5,$ $0.06, 0.08, 0.1, 0.35$	Flux method.	Study of the influence of Ru on the anisotropy of single crystals of PbM. Spin reorientation was observed.	[69]
$PbFe_{12-x}Ga_xO_{19}$ $x = 0, 1, 3, 4,9$	Flux method.	By ellipsometry and polarized light reflectivity measurements, the optical properties of Ga-substituted PbM, like the permittivity, were studied.	[70]
$Pb_aSm_bFe_cTi_dO_z$ $a + b + c + d = 13, Z \approx 19$	Flux method.	Study the influence of donor density on the photo electrochemical properties of single crystals.	[71]
$PbFe_{12-x}Ga_xO_{19}$ $x = 0, 1, 3, 4$	Flux method.	Study of complex polar Kerr effect spectra by diamagnetic substitution in PbM.	[72].
$Ba_{0.8}Pb_{0.2}Fe_{12}O_{19}$ $PbFe_{12}O_{19}$	Flux method and ceramic method.	Measurement of the NMR spectra of ^{57}Fe in domains single crystals and polycrystals of PbM, BaM and SrM.	[73]
$PbFe_{12-x}Cr_xO_{19}$ $0 \leq x \leq 6$	Chemical coprecipitation.	Study of magnetic properties and Mössbauer characterization. A complex magnetic order was induced by Cr^{3+} ions.	[74]

$PbFe_{12-x}Ga_xO_{19}$ $x = 0, 0.5, 1, 1.5$ $PbFe_{12-x}Ga_xO_{19}$ $x = 0, 1, 2$	Metallorganic decomposition.	Study of magnetic properties and Mössbauer characterization. Angled magnetic structures arise due to the substitution	[75]
$PbFe_{12-x}Al_xO_{19}$ $x = 0.0, 3.0, 4.0, 6.0.$	Chemical coprecipitation.	Study of the coercive and magnetization reversal processes by magnetic viscosity measurements.	[76]
$PbFe_{11}CrO_{19}$	Chemical coprecipitation.	Study of reversal and magnetization processes by magnetic viscosity measurements. Two apparent activation volumes associated with different coercivity mechanisms are reported.	[77]
$Sr_{0.5}Pb_{0.5}{}^{2+}Fe_{12-x}Pb_x{}^{3+}O_{19}$ $x = 0.0, 0.2, 0.4, 0.6, 0.8, 1.0$	Ceramic method.	Study of structural and electrical properties. Conductivity is explained by the Verwey model. AC conductivity was explained by the the basis of Maxwell–Wagner model and Koops phenomenological theory.	[78]
$Pb_xBa_{1-x}Fe_{12}CrO_{19}$ $x = 0.1, 0.3, 0.5, 0.7, 0.9$	Ceramic method.	Study of the effect of Ba on magnetic and structural properties of $PbFe_{12}O_{19}$ by Rietveld refinement method, Mossbauer spectroscopy and X-ray diffraction.	[79]

$Sr_{0.5}Pb_{0.5}{}^{2+}Fe_{12-x}Pb_x{}^{3+}O_{19}$ $x = 0.0, 0.2, 0.4, 0.6, 0.8,$ 1.0	Ceramic method.	Study of magnetic properties. Decreasing tendencies in coercivity and remanence are obtained as substitution increases.	[80]
$MFe_{12}O_{19}/KCl$ $MFe_{12}O_{19}/KBr$ $MFe_{12}O_{19}/KI$ $MAlFe_{11}O_{19}/KCl$ $MAlFe_{11}O_{19}/KBr$ $MAlFe_{11}O_{19}/KI$ (M = Sr, Ba and Pb).	Sol-gel method.	Investigated the shielding effect of inorganic ions KCl, KBr and KI on the phase growth of hexaferrites. In the substituted samples, the saturation magnetization decreases because Al^{3+} preferentially occupies the octahedral 2a sites.	[42]
$Pb_{1-x}Sr_xFe_{12}O_{19}$ $x = 0.1, 0.3, 0.5, 0.7$ and 0.9	Ceramic method.	Study of the effect of Sr on structural and magnetic properties of $PbFe_{12}O_{19}$. Extrinsic magnetic properties depend on the strontium substitution. For remanence and saturation the dependence is linear, but for coercivity it is exponential.	[81]

$PbFe_{12-x}Ga_xO_{19}$. $x = 6$.	Flux method.	The magnetic-moment of Fe^{3+} ions at the 2a sublattice of $PbFe_{12-x}Ga_xO_{19}$ is downward while that of PbM is upward at room temperature. The magnetic moment of $PbFe_{12-x}Ga_xO_{19}$ undergoes a reorientation to the upward direction with decreasing temperature while that of the PbM remains unchanged down to 5 K.	[82]
$Ba_{1-x}Pb_xFe_{12}O_{19}$ $x = 0.0, 0.2, 0.4, 0.6, 0.8, 1.0$	Chemical coprecipitation.	Study of the effect of Pb on dielectric and electric properties of nanosized BaM. The resistivity increased and the conductivity decreased with increase in Pb in the Ba–ferrites matrix.	[83]
$Ba_{1-x}Pb_xFe_{12}O_{19}$ $x = 0.0, 0.2, 0.4, 0.6 0.8, 1.0$	Chemical coprecipitation.	Effect of lead on structural and magnetic properties of BaM. The coercivity decreased and the magnetic induction and remanence increased as the concentration of Pb increased.	[84]

$Sr_{1-x}Pb_xFe_{12}O_{19}$ $x = 0.00, 0.05, 0.10, 0.15, 0.20.$	Sol–gel auto-ignition technique	Studied the influence of Pb doping on structural, electrical and magnetic properties of SrM. The increase of Pb decreased dielectric constant and dielectric tangent loss while it decreased saturation magnetization, remanence and increased the coercivity.	[85]
$PbTi_{(1-x)}Fe_xO_{(3-\delta)}$ $x = 0, 0.005, 0.01, 0.025,$ $0.05, 0.075, 0.1, 0.125,$ $0.15, 0.175, 0.2, 0.3.$	Acetic acid based sol–gel route.	Study of the structure of these compounds by the Rietveld refinement method. PbM was observed at compositions $x = 0.4, 0.5, 0.6, 0.7, 0.8, 0.9,$ and 1.0 as a secondary phase.	[86]
$PbFe_6Ga_6O_{19}$	Single crystal.	Study of the relaxor-like ferroelectricity in $PbFe_6Ga_6O_{19}$. It is found that the local spin reversal of Fe^{3+} at the 2a site is a necessary prerequisite for ferroelectricity.	[87]
$PbCo_{0.5}Sn_{0.5}Ho_xFe1_{1-x}O_{19}$ $(x = 0.00–0.75)$	Sol-gel autocombustion method.	Study of the structural, magnetic and microwave absorption properties. The substitution of Ho enhanced the microwave absorption properties of $PbCo_{0.5}Sn_{0.5}Fe_{11}O_{19}$ hexaferrite in the composite ferrite-acrylic resin.	[88]

Eng. Magnetic, Dielectric and Microwave Properties of Ceramics and Alloys   Materials Research Forum LLC
Materials Research Foundations **57** (2019) 23-56   doi: https://doi.org/10.21741/9781644900390-2

$PbCoTiFe_{10}O_{19}$	Ceramic method.	Studied of the multiferroic and magnetoelectric coupling. Direct and converse magnetoelectriceffect are observed in this magnetoelectric hexaferrite.	[89]
$Pb_{1-x}La_xFe_{12}O_{19}$ $x = 0.1, 0.2, 0.3, 0.4, 0.5, 0.7, 0.9$	Ceramic method.	Studied the effect of lanthanum on the structural, morphological, magnetic, and electrical properties of PbM. The fall of two orders of magnitude in the coercivity is associated to the change of valence states of the iron cations located nearby of the lanthanum ions.	[90]
$SrFe_{12}O_{19}$ $Sr_{0.5}Ca_{0.5}Fe_{12}O_{19}$ $Sr_{0.5}Pb_{0.5}Fe_{12}O_{19}$	Ceramic method.	Studied of the structural, dielectric and magnetic properties. The $Sr_{0.5}Pb_{0.5}Fe_{12}O_{19}$ sample showed a reduction in the coercivity and an increase in magnetization with low conductivity, dielectric constant and dielectric loss.	[91]

$Ba_{0.9}Pb_{0.1}Fe_{12}O_{19}$ $Pb_{0.9}Ba_{0.1}Fe_{12}O_{19}$ $Pb_{0.9}Sr_{0.1}Fe_{12}O_{19}$ $PbFe_{12}O_{19}$	Ceramic method.	Studied of the effect of lead on the magnetic interactions in SrM and BaM. With high concentration of lead, demagnetizing interactions prevail and as the lead concentration diminishes the intensity of magnetic interactions also decreases giving rise to exchange interactions.	[92]
$BaFe_{12-x}Pb_xO_{19}$ $x = 0.0, 0.5, 0.75,$ and 1.0	Sol-gel method.	Studied the effect of lead on the structural, morphological, magnetic, and electrical properties. The coercivity and magnetocrystalline anisotropy energy decreased as the substitution of lead increased.	[93]
$Ba_{0.77}Pb_{0.23}Fe_{12}O_{19}$ $Ba_{0.8}Pb_{0.2}Fe_{9.2}Al_{2.8}O_{19}$ $Ba_{0.8}Pb_{0.2}Fe_{8.4}Al_{3.6}O_{19}$ $Ba_{0.84}Pb_{0.16}Fe_{7.18}Al_{4.82}O_{19}$	Flux method.	Studied the influence of Pb and Al on the magnetic properties of BaM. Also, PbO was used as a solvent in order to get a higher level of Al^{3+} substitution, achieving a level of 4.82 in bulk single crystals.	[94]
$La_{0.2}Pb_{0.7}Fe_{12}O_{19}$	Polymer precursor method.	Report the coexistence of large ferroelectricity and strong ferromagnetism and colossal magnetocapacitance at room-temperature.	[95]

$Ba_{1-x}Pb_xFe_{12}O_{19}$ $Ba_{1-x}Pb_xFe_{12}O_{19}$/Polypyrrole $x = 0, 0.1, 0.3, 0.5, 0.7, 0.9, 1$	Sol-gel method.	Investigated the effects of interface of hexaferrite and polypyrrole and the doping of lead on microwave absorption. The reflection loss of ferrite increased in the composite ferrite-PPY and increased as the lead content increased. Also, as x increases the real and imaginary part of the dielectric constant increased. Both effects, content of lead and interface between hexaferrite and PPY are important factors in microwave absorption.	[96]
$PbFe_{12}O_{19}$/CNT $PbFe_{12}O_{19}$/Graphene	Sol-gel auto-combustion route.	Studied the magnetic and electrochemical properties and photocatalytic behavior of PbM. For the first time, the photocatalytic activity of lead hexaferrites is evaluated using the degradation of methyl orange under ultraviolet light irradiation.	[53]
$PbTiO_3$/$PbFe_{12}O_{19}$	Ceramic method.	Obtained better multiferroic properties. Evidence of magneto-electric coupling was found in the composite.	[97]

$La_{0.2}Pb_{0.7}Fe_{12}O_{19}$	Polymer precursor method.	Demonstrated the coexistence of large ferroelectricity, strong ferromagnetism and colossal magnetocapacitance at room temperature in modified PbM.	[98]
$Ba_{0.3}Sr_{0.4}Pb_{0.3}Fe_{12}O_{19}$ $Ba_{0.4}Sr_{0.3}Pb_{0.3}Fe_{12}O_{19}$ $Ba_{0.3}Sr_{0.3}Pb_{0.4}Fe_{12}O_{19}$	Sol-gel autocombustion method.	Study by Mössbauer spectroscopy and measurement of the optical properties of the samples.	[99]
$Ba_{1-x}Pb_xFe_{12}O_{19}$ x = 0.0, 0.1, 0.2, 0.3, 0.4	Chemical coprecipitation method.	Studied the structural, ferroelectric, magnetic and magneto-electric properties of substituted PbM at room temperature. A magneto-electric coupling was observed in these compounds.	[100]
$Ba_{0.5}Sr_{0.5}Fe_{12}O_{19}$ $Ba_{0.4}Sr_{0.4}Pb_{0.2}Fe_{12}O_{19}$ $Ba_{0.3}Sr_{0.3}Pb_{0.4}Fe_{12}O_{19}$ $Ba_{0.4}Sr_{0.3}Pb_{0.3}Fe_{12}O_{19}$ $Ba_{0.3}Sr_{0.4}Pb_{0.3}Fe_{12}O_{19}$	Ceramic method.	Study of the magnetic and microwave properties of Pb-substituted (Ba,Sr)M nanoparticles. The magnetic properties degraded with lead substitution but the absorption of microwaves is improved.	[101]

$PbFe_{12-x}Ga_xO_{19}$ $x = 8.4, 9.0\ 9.6$	Flux method.	The random substitution of Fe by Ga ions for magnetic Fe ions in PbM suppresses the Néel temperature to zero at a critical composition close to the magnetic ion percolation.	[102]

6.   Other studies on PbM

Zainullina et al. made *ab initio* studies with linear muffin-tin orbital method in a tight binding (LMTO-TB) approximation and semi empirical extended Hückel theory (EHT) to study the electronic structure, chemical bonding and ion conductivity of PbM [103]. It was found that the strongest bond corresponds to the Fe-O with the Pb-O bond being insignificant. The magnetic properties were not studied in this work. More recently, the plane-wave pseudopotential density functional theory (DFT) with general gradient approximation (GGA) and GGA + U were used to calculate the structural, electronic and optical properties of pure SrM, BaM, PbM, $Sr_{0.5}Pb_{0.5}Fe_{12}O_{19}$, $Sr_{0.5}Ba_{0.5}Fe_{12}O_{19}$ and $Ba_{0.5}Pb_{0.5}Fe_{12}O_{19}$ [104].

References

[1] Valenzuela, R. Magnetic Ceramics, Cambridge University Press. New York. 1994. https://doi.org/10.1017/CBO9780511600296

[2] Özgür, Ü.,Alivov, Y., Morkoç, H. Microwave ferrites, part 1: fundamental properties. Journal of Materials Science: Materials in Electronics 20 (2009) 789-834. https://doi.org/10.1007/s10854-009-9923-2

[3] Kirchmayr, H. J. Permanent magnets and hard magnetic materials. Journal of Physics D: Applied Physics 29 (1996) 2763–2778. https://doi.org/10.1088/0022-3727/29/11/007

[4] Tan, G., Li, W. Ferroelectricity and ferromagnetism of M-type lead hexaferrite. Journal of the American Ceramic Society 98 (2015) 1812–1817. https://doi.org/10.1111/jace.13530

[5] Kostishin, V. G., Panina, L. V., Kozhitov, L. V., Timofeev, A. V., Zyuzin, A. K., Kovalev, A. N. On synthesis of $BaFe_{12}O_{19}$, $SrFe_{12}O_{19}$, and $PbFe_{12}O_{19}$ hexagonal ferrite ceramics with multiferroic properties. Technical Physics 60 (2015) 1189–1193. https://doi.org/10.1134/S1063784215080150

[6] Toxicological Profile for Lead, US Department of Health and Human Services, Public Health Service, Agency for Toxic Substances and Disease Registry, Atlanta, GA, 1999.

[7] Pullar, R. Hexagonal ferrites: A review of the synthesis, properties and applications of hexaferrite ceramics. Progress in Materials Science 57 (2012) 1191-1334. https://doi.org/10.1016/j.pmatsci.2012.04.001

[8] Mahmood, S. H., Abu-Aljarayesh A. Hexaferrite permanent magnetic materials. Materials Research Foundations, 2016. https://doi.org/10.21741/9781945291074

[9] Adelsköld, V. X-ray studies on magneto-plumbite$PbO(Fe_2O_3)_6$ and other substances resembling beta-alumina $Na_2O(Al_{12}O_3)_{11}$. ArkivförKemi, MineralogiochGeologi, A12 (1938) 1-9.

[10]   Kojima, H. Fundamental properties of hexagonal ferrites with magnetoplumbite structure. In: Wohlfarth E. P., Ed. Ferromagnetic Materials, Vol. 3. Amsterdam: North-Holland Physics Publishing; 1982. https://doi.org/10.1016/S1574-9304(05)80091-4

[11]   Moore, P. B., Sen Gupta, P. K., Le Page, Y. Magnetoplumbite, $Pb^{2+}Fe_{12}{}^{3+}O_{19}$: Refinement and lone-pair splitting. American Mineralogist 74 (1989) 1186-1194.

[12]   Parida, S. C. Order-disorder transitions and thermodynamic properties of M-type hexaferrites. Solid State Phenomena 150 (2009) 101-121. https://doi.org/10.4028/www.scientific.net/SSP.150.101

[13]   Palomares-Sánchez S. A., Díaz-Castañón, S., Ponce-Castañeda S., Mirabal-García, M., Leccabue, F., Watts, B. E. Use of the Rietveld refinement method for the preparation of pure lead hexaferrite. Material Letters 59 (2005) 591-594. https://doi.org/10.1016/j.matlet.2004.11.002

[14]   Panda, A., Govindaraj, R., Vinod, K., Kalavathi, S., Amarendra, G. Nanoparticles of lead hexaferrite as studied using Mössbauer spectroscopy. AIP Conference Proceedings 1832 (2017) 050153-1-3. https://doi.org/10.1063/1.4980386

[15]   Watts, B. E., Regonini, D., Leccabue, F., Casoli, F., Albertini, F., Bocelli, G., Schmool, D. Structural and magnetic properties of chemically deposited hexaferrites. Materials Science Forum 514-516 (2006) 304-308. https://doi.org/10.4028/www.scientific.net/MSF.514-516.304

[16]   Ikeda, Y., Hara, C., Fujii, T., Sato, M., Inoue, M. Direct synthesis of lead-hexaferrite particles by mist pyrolysis. Journal of the Magnetics Society of Japan 22 (1998) 249–251. https://doi.org/10.3379/jmsjmag.22.S1_249

Eng. Magnetic, Dielectric and Microwave Properties of Ceramics and Alloys   Materials Research Forum LLC
Materials Research Foundations **57** (2019) 23-56   doi: https://doi.org/10.21741/9781644900390-2

[17]   Xu, P., Han, X., Wang, M. Synthesis and magnetic properties of $BaFe_{12}O_{19}$ hexaferrite nanoparticles by a reverse microemulsion technique. Journal of Physical Chemistry C 111 (2007) 5866–5870. https://doi.org/10.1021/jp068955c

[18]   Cabañas, M. V., González-Calbet, J. M., Labeau, M., Mollard, P., Pernet, M., Vallet-Regi, M. Evolution of the microstructure and its influence on the magnetic properties of aerosol synthesized $BaFe_{12}O_{19}$ particles. Journal of Solid State Chemistry 101 (1992) 265-274. https://doi.org/10.1016/0022-4596(92)90183-V

[19]   Ramírez, A. E., Solarte, N. J., Singh, L. H., Coaquirac, J. A. H., Gaona J., S. Investigation of the magnetic properties of $SrFe_{12}O_{19}$ synthesized by the Pechini and combustion methods. Journal of Magnetism and Magnetic Materials 438 (2017) 100-106. https://doi.org/10.1016/j.jmmm.2017.04.042

[20]   Schumacher, F., Hempel, K., Von Staa, F. The magnetic behavior of barium ferrite prepared by glass crystallization method. Journal de Physique Colloques 49 (1988) C8-949-C8-950. https://doi.org/10.1051/jphyscol:19888433

[21]   Matirosyan, K. S., Matirosyan, N., Chalykh, A. Structure and properties of hard–magnetic barium, strontium, and lead ferrites. Inorganic Materials 39 (2003) 886-870.

[22]   Gajbhiye, N. S., Vijayalakshmi, Weissmüller, A. Magnetic properties of nanosize lead hexaferrite particles. physica status solidi 189 (2002) 685-689. https://doi.org/10.1002/1521-396X(200202)189:3<685::AID-PSSA685>3.0.CO;2-S

[23]   Korniyenko K. Iron – Oxygen – Lead. In: Effenberg G., Ilyenko S. (eds) Ternary Alloy Systems. Landolt-Börnstein - Group IV Physical Chemistry. Vol 11D5 (2009) Springer, Berlin. https://doi.org/10.1007/978-3-540-70890-2_19

[24]   Berger, W., Pawlek, F. Kristallographische und magnetische Untersuchungen im System Bleioxyd (PbO)-Eisenoxyd (Fe_2O_3). Archivfür das Eisenhüttenwesen 28 (1957)101-108. https://doi.org/10.1002/srin.195702122

[25]   Mountvala, A. J., Ravitz, S. F. Phase relations and structures in the system PbO-Fe_2O_3. Journal of the American Ceramic Society. 45 (1962) 285-288. https://doi.org/10.1111/j.1151-2916.1962.tb11146.x

[26]   Nevřiva, M., Fischer, K. Contribution to the binary phase diagram of the system PbO-Fe_2O_3. Materials Research Bulletin 21 (1986) 1285-1290. https://doi.org/10.1016/0025-5408(86)90061-9

[27]   Cocco, A. Richerche sul sistema binario PbO- Fe_2O_3.Annali di Chimica (Roma) 45 (1955) 737-53.

Eng. Magnetic, Dielectric and Microwave Properties of Ceramics and Alloys   Materials Research Forum LLC
Materials Research Foundations **57** (2019) 23-56   doi: https://doi.org/10.21741/9781644900390-2

[28]   Tokar, M. Microstructure and magnetic properties of lead ferrite. Journal of the American Ceramic Society 52 (1968) 302-306. https://doi.org/10.1111/j.1151-2916.1969.tb11930.x

[29]   Shono, K., Gomi, M., Abe, M. Magneto-optical properties of magnetoplumbites $BaFe_{12}O_{19}$, $SrFe_{12-x}Al_xO_{19}$ and $PbFe_{12}O_{19}$. Japanese Journal of Applied Physics 21 (1982) 1451-1454. https://doi.org/10.1143/JJAP.21.1451

[30]   Ram, S., Bahadur, D., Chakravorty, D. Magnetic glass-ceramics with hexagonal lead ferrites. Journal of Non-Crystalline Solids 88 (1986) 311-322. https://doi.org/10.1016/S0022-3093(86)80033-3

[31]   Višňovský, Š.,Široký, P. Krishnan, R. Complex polar Kerr effect spectra of magnetoplumbite. Czechoslovak Journal of Physics B 36 (1986) 1434–1442. https://doi.org/10.1007/BF01959568

[32]   Štěpánková, H., Englich, J., Lütgemeier H. NMR study of Ga and Al substituted hexagonal ferrites with magnetoplumbite structure. IEEE Transactions on Magnetics 30 (1994) 988-990. https://doi.org/10.1109/20.312467

[33]   Carp, O., Segal, E., Brezeanu, M., Barjega, Stanica, N. Nonconventional methods for obtaining hexaferrites. I. Lead hexaferrite. Journal of Thermal Analysis 50 (1997) 125–135. https://doi.org/10.1007/BF01979555

[34]   Bezlepkin, A. A., Kuntsevich, S. P., Kostyukov, V. I. Magnetic relaxation of oscillating domain walls in $PbFe_{12}O_{19}$. Physics of the Solid State 39 (1997) 99-100. https://doi.org/10.1134/1.1129840

[35]   Ikeda, Y., Hara, C., Fujii, T., Sato, M., Inoue, M. Direct synthesis of lead-hexaferrite particles by mist pyrolysis. Journal of the Magnetics Society of Japan 22 (1998) 249–251. https://doi.org/10.3379/jmsjmag.22.S1_249

[36]   Díaz-Castañón S., Sánchez Ll, J. L., Estevez-Rams, E., Leccabue, F., Watts, B. Magneto-structural properties of $PbFe_{12}O_{19}$ hexaferrite powders prepared by decomposition of hydroxide–carbonate and metal–organic precipitates. Journal of Magnetism and Magnetic Materials 185 (1998) 194-198. https://doi.org/10.1016/S0304-8853(98)00013-4

[37]   Zhukovsky, V. M., Bushkova, O. V., Zainullina, V. M., Dontsov, G. I., Volosentseva, L. I., Zhukovskaya A. S. Diffusion transport in hexagonal ferrites with magnetoplumbite structure. Solid State Ionics 119 (1999) 15-17. https://doi.org/10.1016/S0167-2738(98)00476-7

[38]   Gajbhiye, N. S., Vijayalakshmi, Weissmüller, A. Magnetic properties of nanosize lead hexaferrite particles. physica status solidi 189 (2002) 685-689. https://doi.org/10.1002/1521-396X(200202)189:3<685::AID-PSSA685>3.0.CO;2-S

[39]   Matirosyan, K. S., Matirosyan, N., Chalykh, A. Structure and properties of hard–magnetic barium, strontium, and lead ferrits. Inorganic Materials 39 (2003) 886-870. https://doi.org/10.1023/A:1025037716108

[40]   Díaz-Castañón, S., Faloh-Gandarilla, J., Leccabue F., Albanese G. The optimum synthesis of high coercivity Pb-M hexaferrite powders using modifications to the traditional ceramic route. Journal of Magnetism and Magnetic Materials, 272-276 (2004) 2221-2223. https://doi.org/10.1016/j.jmmm.2003.12.923

[41]   Yang, N., Yang, H., Jia, J., Pang, X. Formation and magnetic properties of nanosized $PbFe_{12}O_{19}$ particles synthesized by citrate precursor technique. Journal of Alloys and Compounds 438 (2007) 263-267. https://doi.org/10.1016/j.jallcom.2006.08.037

[42]   Chaudhury, S., Rakshit, S. K., Parida, S. C., Singh, Z., Singh Mudher, K. D., Venugopal, V. Studies on structural and thermo-chemical behavior of $MFe_{12}O_{19}(s)$ (M = Sr, Ba and Pb) prepared by citrate-nitrate gel combustion method. Journal of Alloys and Compounds 455 (2008) 25-30. https://doi.org/10.1016/j.jallcom.2007.01.075

[43]   Singhal, S., Namgyal, T., Singh, J., Chandra, K., Bansal, S. A comparative study on the magnetic properties of $MFe_{12}O_{19}$ and $MAlFe_{11}O_{19}$ (M = Sr, Ba and Pb) hexaferrites with different morphologies. Ceramics International 37 (2011) 1833-1837. https://doi.org/10.1016/j.ceramint.2011.02.001

[44]   Tan, G., Wang, M. Multiferroic $PbFe_{12}O_{19}$ceramics. Journal of Electroceramics 26 (2011) 170-174. https://doi.org/10.1007/s10832-011-9641-z

[45]   Mao, L., Cui, H., An, H,, Wang, B., Zhai, J., Zhao, Y., Li, Q. Stabilization of simulated lead sludge with iron sludge via formation of $PbFe_{12}O_{19}$ by thermal treatment. Chemosphere 117 (2014) 745-752. https://doi.org/10.1016/j.chemosphere.2014.08.027

[46]   Ansari, F., Sobhani, B., Salavati-Niasari, M. Sol–gel auto-combustion synthesis of $PbFe_{12}O_{19}$ using maltose as a novel reductant. RSC Advances 4 (2014) 63946-63950. https://doi.org/10.1039/C4RA11688G

[47]   Guerrero-Serrano, A. L.; Mirabal-García, M.; Palomares-Sánchez, S. A.; Martínez-Mendoza, J. R. Study of the magnetic properties of the Pb-hexaferrite obtained as a single phase by two methods. RevistaLatinoamericana de Metalurgia y Materiales 34 (2014) 136-141.

[48]   Ansari, F., Soofivand, F., Salavati-Niasari, M. Utilizing maleic acid as a novel fuel for synthesis of $PbFe_{12}O_{19}$nanoceramics via sol-gel auto-combustion route. Materials Characterization 103 (2015) 11-17. https://doi.org/10.1016/j.matchar.2015.03.010

[49]   Ansari, F., Salavati-Niasari, M. Simple sol-gel auto-combustion synthesis and characterization of lead hexaferrite by utilizing cherry juice as a novel fuel and green capping agent. Advanced Powder Technology 27 (2016) 2025-2031. https://doi.org/10.1016/j.apt.2016.07.011

[50]   Halakouie, H., Nabiyouni, G., Saffari, J., Ahmadi, A., Ghanbari, D. Lead hexa-ferrites and magnetic cellulose acetate nanocomposites: study of magnetization, coercivity and remanence. Journal of Materials Science: Materials in Electronics 27 (2016) 7738-7749. https://doi.org/10.1007/s10854-016-4761-5

[51]   MousaviGhahfarokhi, S. E., Rostami, Z., Kazeminezhad, I. Fabrication of $PbFe_{12}O_{19}$ nanoparticles and study of their structural, magnetic and dielectric properties. Journal of Magnetism and Magnetic Materials 399 (2016) 130-142. https://doi.org/10.1016/j.jmmm.2015.09.063

[52]   Ansari, F., Sobhani, A., Salavati-Niasari, M. $PbTiO_3/PbFe_{12}O_{19}$ nanocomposites: Green synthesis through an eco-friendly approach. Composites Part B 85 (2016) 170-175. https://doi.org/10.1016/j.compositesb.2015.09.027

[53]   Asiabani, N., Nabiyouni, G., Khaghani, S., Ghanbari, D. Green synthesis of magnetic and photo-catalyst $PbFe_{12}O_{19}$-PbS nanocomposites by lemon extract: nano-sphere $PbFe_{12}O_{19}$ and star-like PbS. Journal of Materials Science: Materials in Electronics 28 (2017) 1101-1114. https://doi.org/10.1007/s10854-016-5635-6

[54]   Mahdiani, M., Sobhani, A., Salavati-Niasari, M. Enhancement of magnetic, electrochemical and photocatalytic properties of lead hexaferrites with coating graphene and CNT: Sol-gel auto-combustion synthesis by valine. Separation and Purification Technology 185 (2017) 140-148. https://doi.org/10.1016/j.seppur.2017.05.029

[55]   Prathap, S., Madhuri, W. Multiferroic properties of microwave sintered $PbFe_{12-x}O_{19-\delta}$. Journal of Magnetism and Magnetic Materials 430 (2017) 114–122. https://doi.org/10.1016/j.jmmm.2016.12.116

[56]   Lu, X., Ning, X., Lee, P.-.H, Shih, K., Wang, F., Zeng, E. Y. Transformation of hazardous lead into lead ferrite ceramics: Crystal structures and their role in lead leaching. Journal of Hazardous Materials 336 (2017) 139–145. https://doi.org/10.1016/j.jhazmat.2017.04.061

[57]   Mahdiani, M., Soofivand, F., Salavati-Niasari, M. Investigation of experimental and instrumental parameters on properties of $PbFe_{12}O_{19}$ nanostructures prepared by sonochemical method. UltrasonicsSonochemistry 40 (2017) 271-281. https://doi.org/10.1016/j.ultsonch.2017.06.023

[58]   Morisako, A., Nakanishi, H., Matsumoto, M., Naoe, M. Low-temperature deposition of hexagonal ferrite films by sputtering. Journal of Applied Physics 75 (1994) 5969-5971. https://doi.org/10.1063/1.355528

[59]   Dorsey P. C., Qadri S. B., Grabowski, K. S., Knies, D. L., Lubitz, P., Chrisey, D. B., Horwitz, J. S. Epitaxial Pb–Fe–O film with large planar magnetic anisotropy on (0001) sapphire. Physical Review Letters 70 (1997) 1173-1175. https://doi.org/10.1063/1.118483

[60]   Díaz-Castañón, S., Leccabue, F., Watts, B. E., Yapp, R., Asenjo, A., Vázquez, M. Oriented $PbFe_{12}O_{19}$ thin films prepared by pulsed laser deposition on sapphire substrate. Materials Letters 47 (2001) 356-361. https://doi.org/10.1016/S0167-577X(00)00266-4

[61]   Díaz-Castañón, S., Leccabue, F., Watts, B. E., Yapp, R. $PbFe_{12}O_{19}$ thin films prepared by pulsed laser deposition on Si/SiO_2 substrates. Journal of Magnetism and Magnetic Materials 220 (2000) 79-84. https://doi.org/10.1016/S0304-8853(00)00473-X

[62]   Faloh-Gandarilla, J. C., Díaz-Castañón S., Leccabue, F., Watts, B. E. Magnetic properties of polycrystalline Sr-M and Pb-M hexaferrites thin films grown by pulsed laser deposition on Si/SiO_2 substrates. Journal of Alloys and Compounds 369 (2004) 195-197. https://doi.org/10.1016/j.jallcom.2003.09.102

[63]   Castro-Rodríguez, R., Palomares-Sánchez, S., Leccabue, F., Arisi, E., Watts, B. E. Optimal target-substrate distance in the growth of oxides thin films by pulsed laser deposition. Materials Letters 57 (2003) 3320-3324. https://doi.org/10.1016/S0167-577X(03)00066-1

[64]   Doh, S. J., Je, J. H., Cho, T. S. Pb cation induced low-temperature crystallization of (Ba·Pb) hexa-ferrite thin films. Journal of Electroceramics 17 (2006) 365-368. https://doi.org/10.1007/s10832-006-7239-7

[65]   Geiler, A. L., He, Y., Yoon, S. D., Yang, A., Chen, Y., Harris, V. G., Vittoria, C. Epitaxial growth of $PbFe_{12}O_{19}$ thin films by alternating target laser ablation deposition of Fe_2O_3 and PbO. Journal of Applied Physics 101 (2007) 09M510-1-3. https://doi.org/10.1063/1.2710222

[66]   Croft, W. J., Kestigian, M., Borovicka, R., Garabedian, F. Unit cell dimensions in the system $PbAl_{12-x}Fe_xO_{19}$. Materials Research Bulletin 2 (1967) 849-852. https://doi.org/10.1016/0025-5408(67)90093-1

[67]   Pollert, E., Nevřiva, M., Matějková, L., Novák, J. Preparation and characterization of $PbFe_{12-x}Ga_xO_{19}$ single crystals. Materials Research Bulletin 16 (1981) 1499-1504. https://doi.org/10.1016/0025-5408(81)90020-9

[68]   Pollert, E., Matějková, L. Single crystals of lead hexaferrite substituted by samarium ions. Crystal Research and Technology 16 (1981) K53-K54. https://doi.org/10.1002/crat.19810160327

[69]   Zhai, H. R., Liu, J. Z., Lu, M. Influence of Ru^{3+} ions on anisotropy of $PbFe_{12}O_{19}$ single crystals. Journal of Applied Physics 52 (1981) 2323-2325. https://doi.org/10.1063/1.328919

[70]   Široký, P., Schmidt, E., Lukeš, F., Humlíček, J. Optical properties of Ga-substituted magnetoplumbites.physical status solidi 83 (1984) 581-588. https://doi.org/10.1002/pssa.2210830220

[71]   Pollert, E., Hejtmanek, J., Doumerc, J. P. Influence of the donor density on the photoelectrochemical properties of the magnetoplumbite $PbFe_{12}O_{19}$. Journal of Inorganic and General Chemistry540 (1986) 205-211. https://doi.org/10.1002/chin.198707017

[72]   Široký, P. Višňovský, Š. Magneto-optical properties of Ga-substituted magnetoplumbites. Czechoslovak Journal of Physics B 37 (1987) 116-121. https://doi.org/10.1007/BF01597886

[73]   Štěpánková, H., Englich, J., Novák, P., Sedlák, B., Pfeffer, M. NMR spectra of ^{57}Fe in hexagonal ferrites with magnetoplumbite structure. Hyperfine Interactions 50 (1989) 639-643. https://doi.org/10.1007/BF02407702

[74]   Albanese, G., Watts, B. E,,Leccabue, F., Díaz-Castañón, S. Mössbauer and magnetic studies of $PbFe_{12-x}Cr_xO_{19}$ hexagonal ferrites. Journal of Magnetism and Magnetic Materials 184 (1998) 337-343. https://doi.org/10.1016/S0304-8853(97)01162-1

[75]   Albanese, G., Díaz-Castañón, S. Leccabue, F., Watts, B. E. Mössbauer and magnetic investigation of scandium and indium substituted $PbFe_{12}O_{19}$ hexagonal ferrite. Journal of Materials Science 35 (2000) 4415-4420. https://doi.org/10.1023/A:1004869310024

[76]   FalohGandarilla J. C., Díaz-Castañón, S,,SuárezAlmodovar, N. Activation volume and coercivity in aluminum-substituted Pb-M hexaferrites. Journal of

Magnetism and Magnetic Materials 222 (2000) 271-276.
https://doi.org/10.1016/S0304-8853(00)00431-5

[77]   Faloh-Gandarilla J, C., Díaz-Castañón, S., Leccabue, F. Magnetic viscosity and activation volume in chromium substituted Pb-M hexaferrite. physica status solidi (B) Basic Research 242 (2005) 1784-1787. https://doi.org/10.1002/pssb.200461836

[78]   Hussain, S., Maqsood, A. Structural and electrical properties of Pb-doped Sr-hexa ferrites. Journal of Alloys and Compounds 466 (2008) 293-298.
https://doi.org/10.1016/j.jallcom.2007.11.074

[79]   Guerrero-Serrano, A, L. Pérez-Juache ,T., Mirabal-García, M., Matutes-Aquino, J. A., Palomares-Sánchez, S. A. Effect of barium on the properties of lead hexaferrite. Journal of Superconductivity and Novel Magnetism 24 (2011) 2307-2312.
https://doi.org/10.1007/s10948-011-1181-x

[80]   Hussain, S., Shah, N. A., Maqsood, A., Ali, A., Naeem, M., Syed W. A. A. Characterization of Pb-doped Sr-ferrites at room temperature. Journal of Superconductivity and Novel Magnetism24 (2011) 1245-1248.
https://doi.org/10.1007/s10948-010-1115-z

[81]   Guerrero-Serrano, A, L., Palomares-Sánchez, S. A., Mirabal-García, M., Matutes-Aquino, J. A. Magneto-structural characterization of strontium substituted lead hexaferrite. Journal of Superconductivity and Novel Magnetism 25 (2012) 1223-1228. https://doi.org/10.1007/s10948-012-1411-x

[82]   Na, E. H., Lee, J.-H., Ahn, S.-J., Hon, K.-P., Koo, Y. M., Jang, H. M. Local spin reversal and associated magnetic responses in Ga-substituted Pb-hexaferrites. Journal of Magnetism and Magnetic Materials 324 (2012) 2866-2870.
https://doi.org/10.1016/j.jmmm.2012.04.031

[83]   Haq, A., Anis-ur-Rehman, M., Malik, M. A. Structural and electrical transport properties of proficient Ba–Pbnanoferrites. PhysicaScripta 85 (2012) 035602.
https://doi.org/10.1088/0031-8949/85/03/035602

[84]   Haq, A., Anis-ur-Rehman, M., Malik, M. A. Effect of Pb on structural and magnetic properties of Ba-hexaferrite. Physica B 407 (2012) 822-826.
https://doi.org/10.1016/j.physb.2011.11.038

[85]   Ullah, Z., Atiq, S., Naseem, S. Influence of Pb doping on structural, electrical and magnetic properties of Sr-hexaferrites. Journal of Alloys and Compounds 555 (2013) 263-267. https://doi.org/10.1016/j.jallcom.2012.12.061

[86]   Ganegoda, H., Kaduk, J. A., Segre, C. U. X-ray powder diffraction refinement of $PbTi_{(1-x)}Fe_xO_{(3-\delta)}$ solid solution series. Powder Diffraction 28 (2013) 238-245. https://doi.org/10.1017/S0885715613000511

[87]   Na, E. H., Song, S., Koo, Y.-M., Jang, H. Relaxor-like improper ferroelectricity induced by $S_i \cdot S_j$-type collinear spin ordering in a M-type hexaferrite $PbFe_6Ga_6O_{19}$. ActaMaterialia 61 (2013) 7705-7711. https://doi.org/10.1016/j.actamat.2013.09.007

[88]   Sharbati, A., Mola, J., Khani, J. Influence of Ho substitution on structural, magnetic and microwave absorption properties of PbM-type hexaferrites nanoparticles. Journal of Materials Science: Materials in Electronics 25 (2014) 244-248. https://doi.org/10.1007/s10854-013-1578-3

[89]   Zhou, W. P., Wang, L. Y., Song, Y. Q., Fang, Y. Q., Wang, D. H., Cao, Q. Q., Du, Y. W. Magnetoelectric effect in $PbCoTiFe_{10}O_{19}$ multiferroic ceramic. Ceramics International 40 (2014) 15737-15742. https://doi.org/10.1016/j.ceramint.2014.07.097

[90]   Guerrero-Serrano, A. L., Mirabal-García, M., Palomares-Sánchez, S. A. Synthesis and study of the lanthanum substitution in the lead M-type hexaferrite. Journal of Superconductivity and Novel Magnetism 27 (2014) 1709-1713. https://doi.org/10.1007/s10948-014-2489-0

[91]   Hooda, A., Sanghi, S., Agarwal, A., Dahiya, R. Crystal structure refinement, dielectric and magnetic properties of Ca/Pb substituted $SrFe_{12}O_{19}$ hexaferrites. Journal of Magnetism and Magnetic Materials 387 (2015) 46-51. https://doi.org/10.1016/j.jmmm.2015.03.078

[92]   Guerrero, A. L. Mirabal-García, M., Palomares-Sánchez, S. A., Martínez, J. R. Effect of Pb on the magnetic interactions of the M-type hexaferrites. Journal of Magnetism and Magnetic Materials 399 (2016) 41-45. https://doi.org/10.1016/j.jmmm.2015.09.052

[93]   Haq, A., Tufail, M., Anis-ur-Rehman, M. Structural, electrical, and magnetic properties of $BaFe_{12-x}Pb_xO_{19}$ hexaferrite. Journal of Superconductivity and Novel Magnetism (2016). In Press.

[94]   Vinnik, D. A., Gudkova, S. A., Niewa, R. Growth of lead and aluminum substituted barium hexaferrite single crystals from lead oxide flux. Materials Science Forum 843 (2016) 3-9. https://doi.org/10.4028/www.scientific.net/MSF.843.3

[95]   Tan, G.-L., Sheng, H.-H. Multiferroism and colossal magneto-capacitance effect of $La_{0.2}Pb_{0.7}Fe_{12}O_{19}$ ceramics. Acta Materialia 121 (2016) 144-151. https://doi.org/10.1016/j.actamat.2016.08.083

[96]   Jin, J., Liu, Y., Drew, M. G. B., Liu, Y. Preparation and characterizations of $Ba_{1-x}Pb_xFe_{12}O_{19}$/polypyrrole composites. Journal of Materials Science: Materials in Electronics 28 (2017) 11325–11331. https://doi.org/10.1007/s10854-017-6925-3

[97]   Jaffari, G. H., Bilal, M., Rahman, J. U., Lee, S. Formation of multiferroic. $PbTiO_3$/$PbFe_{12}O_{19}$ composite by exceeding the solubility limit of Fe in $PbTiO_3$.Physica B: Condensed Matter 520 (2017) 139-147. https://doi.org/10.1016/j.physb.2017.06.035

[98]   Tan, G.-L., Sheng, H.-H. Multiferroic $La_{0.2}Pb_{0.7}Fe_{12}O_{19}$ ceramics: Ferroelectricity, ferromagnetism and colossal magneto-capacitance effect. Data in Brief 10 (2017) 69-74. https://doi.org/10.1016/j.dib.2016.11.067

[99]   Baykal, A., Yokuş, S., Güner, S., Güngüneş, H., Sözeri, H., Amir, Md. Magneto-optical properties and Mössbauer investigation of $Ba_xSr_yPb_zFe_{12}O_{19}$ hexaferrites. Ceramics International 43 (2017) 3475-3482. https://doi.org/10.1016/j.ceramint.2016.10.013

[100]   Kumar, P., Gaur, A., Kotnala, R. Magneto-electric response in Pb substituted M-type barium-hexaferrite. Ceramics International 43 (2017) 1180-1185. https://doi.org/10.1016/j.ceramint.2016.10.060

[101]   Baykal, A., Ünver, İ. S., Topal, U., Sözeri, H. Pb substituted Ba,Sr-hexaferrite nanoparticles as high quality microwave absorbers. Ceramics International (2017). In Press. https://doi.org/10.1016/j.ceramint.2017.07.134

[102]   Rowley, S. E, Vojta, T., Jones, A. T., guo, W., Oliveira, J., Morrison, F. D., Lindfield, N., BaggioSaitovitch, E., Watts, B. E., Scott, J. F. Quantum percolation phase transition and magnetoelectric dipole glass in hexagonal ferrites. Physical Review B 96 (2017) 020407. https://doi.org/10.1103/PhysRevB.96.020407

[103]   Zainullina, V., Zhukov, V., Zhukovskii. V. Quantum-chemical calculation of the electronic structure and ionic conductivity of lead hexaferrite with a magnetoplumbite structure. Journal of Structural Chemistry 42 (2001) 705-710. https://doi.org/10.1023/A:1017948812398

[104]   Sun, W., Zhang, L., Liu, J., Wang, H., Zuo, Y., Bu, Y. First-principle study of the electronic structures and optical properties of six typical hexaferrites. Computational Materials Science 105 (2015) 27-31. https://doi.org/10.1016/j.commatsci.2015.04.021

Eng. Magnetic, Dielectric and Microwave Properties of Ceramics and Alloys   Materials Research Forum LLC
Materials Research Foundations **57** (2019) 57-74   doi: https://doi.org/10.21741/9781644900390-3

Chapter 3

Crystal Structure and Functional Properties of Ni-Fe Films

A.V. Trukhanov[1, 2], S.S. Grabchikov[1], S.V. Trukhanov[1, 2*], A.A. Solobai[1], V.A. Turchenko[3], E.L. Trukhanova[1], T.I. Zubar[1,2], D.I. Tishkevich[1,2], D.A. Vinnik[2]

[1]SSPA "Scientific and practical materials research centre of NAS of Belarus", P. Brovki str., 19, Minsk, 220072 Belarus, *e-mail: sv_truhanov@mail.ru

[2]South Ural State University, 454080, Chelyabinsk, Lenin Prospect, 76,Russia

[3]Joint institute for nuclear research, J. Cuire, 6, Dubna, Moscow region, 141980 Russia

Abstract

$Ni_{1-x}Fe_x$ alloys were produced in film form via electrodeposition. These samples were produced to study magnetic characteristics as a function of chemical composition $Ni_{1-x}Fe_x$ with x = 0; 0.20 and 0.50 and film thickness for the $Ni_{80}Fe_{20}$ system. The chemical composition corresponds to that established before synthesis. Deviation from the calculated stoichiometry is less than 0.7 at. %. The main magnetic parameters of obtained films such as permeability, coercivity and induction were investigated as a function of Fe concentration. The permeability and coercivity as a function of the thickness for the $Ni_{80}Fe_{20}$ films were measured. The field dependence of initial permeability for these films was also measured in the thickness range from 100 nm to 80 μm.

Keywords

NiFe Alloys, Permalloy Films, Crystal Structure, Magnetic Properties

Contents

Eng. Magnetic, Dielectric and Microwave Properties of Ceramics and Alloys   Materials Research Forum LLC
Materials Research Foundations **57** (2019) 57-74   doi: https://doi.org/10.21741/9781644900390-3

1.   Introduction

NiFe alloys have attracted much attention [1-5] and are widely used for practical applications such as magnetic recording media, sensors, spintronic materials and magnetostatic shields. NiFe alloys or permalloys are separated into two large classes – RICH permalloy satNi content >70 at.% and POOR permalloy satNi content ≤50 at.%. These materials have optimal magnetic such as high permeability, low coercivity, and small magnetic anisotropy and functional such as a magnetoresistance and magnetostatic shielding properties [1-8].

Protection from magnetic fields is important [6-10]. There are two ways to protect electronics from external fields: active protection and passive shielding. Active protection is based on conductive wire coils that produce oppositely directed magnetic fields. The resulting field, which is a result of the superposition of external field and the opposite field, is lower than the external field. However, this approach has many disadvantages. The coil design is very expensive and must be correctly oriented to external fields and the protected area. Passive shielding is based on local shields that are produced on the housing surfaces of the protected area. Magnetostatic shields are usually made of ferromagnetic materials with high permeability [6]. Ferromagnetic materials reduce the magnetic flux density in a shielded area through the flux line deviation [11, 12]. Materials with high electrical conductivity are sometimes used for passive shielding [13, 14]. Conductive shields are made of copper or aluminum, and the latter is often preferred because of its lower specific weight. The working principle of conductive shields is based on Faraday's law, but conductive materials are suitable for practical applications only and need to be protected from alternating magnetic fields. The shielding effectiveness of conducting non-magnetic materials increases with increasing radiation frequency [15, 16]. For permanent or extremely low frequency fields, these materials are quite ineffective.

In this paper, we describe the features of $Ni_{1-x}Fe_x$ ($0 \leq x \leq 0.5$) film production. We did experimental studies of the crystal structure parameters and magnetic properties for $Ni_{1-x}Fe_x$ ($0 \leq x \leq 0.5$) films as a function of chemical composition. We measured the magnetic

Eng. Magnetic, Dielectric and Microwave Properties of Ceramics and Alloys   Materials Research Forum LLC
Materials Research Foundations **57** (2019) 57-74   doi: https://doi.org/10.21741/9781644900390-3

properties for $Ni_{80}Fe_{20}$ films as a function of the film thickness in the range of 100 nm – 80 μm. This information will be critically important in developing highly efficient shields against permanent magnetic field [17]. These shields can be used for protection different device bodies with wide-ranging applications including studying the influence of external permanent magnetic fields.

2.   Experimental

The $Ni_{1-x}Fe_x$ (0<x<0.5) films as well as $Ni_{80}Fe_{20}$/Cu multilayered structures were produced via electrodeposition [18]. Aluminium foil was used as the substrate for $Ni_{1-x}Fe_x$ (0<x<0.5) film production. The thickness of the $Ni_{1-x}Fe_x$ (0<x<0.5) films was 50 μm. Three Ni, $Ni_{80}Fe_{20}$ and $Ni_{50}Fe_{50}$ samples were used to determine the correlation between chemical composition, crystal structure parameters and main magnetic characteristics such as a magnetic permeability – μ, coercivity and induction.

The $Ni_{80}Fe_{20}$ films were produced on an Al-substrate with different thicknesses from 100 nm-80 μm. These were used to study the main magnetic parameters as a function of film thickness.

We produced five multilayered periodic structures, i.e. the magnetostatic shields to measure the shielding effectiveness. The aluminum billets were cylindrically shaped with an external diameter of 60 mm, an inner diameter of 56 mm and a length of 100 mm. These were used as the substrates for the $Ni_{80}Fe_{20}$/Cu multilayered structures, i.e. the shields. The total thickness of the magnetic layers of the shields was held constant at 400 μm. The number and thickness of the partial magnetic layer varied. The thickness of the partial Cu layer was fixed at 2 μm: 1). 1x400 $\mu mNi_{80}Fe_{20}$; 2). 5x80 μm $Ni_{80}Fe_{20}$/4x2 μm Cu; 3). 10x40 μm $Ni_{80}Fe_{20}$/9x2 μm Cu; 4). 40x10 μm $Ni_{80}Fe_{20}$/39x2 μm Cu and 5). 80x5 μm $Ni_{80}Fe_{20}$/79x2 μm Cu

The magnetic film (Ni) was formed via electrodeposition from a combined electrolyte: $NiSO_4$ – 280 g/l, $NiCl_2$ – 25 g/l, H_3BO_3 – 30 g/l, $MgSO_4$ – 60 g/l, saccharin– 1 g/l. Parameters of electrodeposition: pH = 2.3–2.5, D_k = 20–30 mA/cm^2, T = 32–37°C.

The magnetic film ($Ni_{50}Fe_{50}$) was formed via electrodeposition from the combined electrolyte: $NiSO_4$ – 200 g/l, $NiCl_2$ – 12 g/l, H_3BO_3 – 30 g/l, $MgSO_4$ – 60 g/l, $FeSO_4$ –71 g/l and saccharin– 2 g/l. Parameters of electrodeposition: pH = 2.4–2.6, $D_к$ = 27–34 mA/cm^2, T = 28–31°C.

The magnetic films ($Ni_{80}Fe_{20}$) were formed via electrodeposition from the combined electrolyte: $NiSO_4$– 210 g/l, $NiCl_2$– 20 g/l, H_3BO_3 – 30 g/l, $MgSO_4$ – 60 g/l, $FeSO_4$ –15 g/l, saccharin– 1 g/l. Parameters of electrodeposition: pH = 2.3–2.5, $D_к$ = 20–30 mA/cm^2,

Eng. Magnetic, Dielectric and Microwave Properties of Ceramics and Alloys   Materials Research Forum LLC
Materials Research Foundations **57** (2019) 57-74   doi: https://doi.org/10.21741/9781644900390-3

T = 30–35°C. The film thicknesses were 100 nm to 80 μm (for investigation of the main magnetic parameters as a function of the film thickness) and 5 to 400 μm (for partial magnetic layer in multilayered shields). The number of magnetic layers in the shields varied from 1 to 80.

Conductive layers (Cu) were formed via electrodeposition from the electrolyte: $CuSO_4 \cdot 5H_2O$ – 35-40 g/l; potassium pyrophosphate – 145 g/l; $KNaC_4H_4O_6$ $4H_2O$ – 25 g/l; $NaHPO_4$ – 45 g/l. Parameters of electrodeposition: pH = 8.2–8.4 D_κ = 5–7 mA/cm^2, T = 35–40°C. The thickness of the partial layers is 2μm. The number of copper layers in the shields is varied from 4 to 79.

The chemical composition of the $Ni_{1-x}Fe_x$ ($0 \leq x \leq 0.5$) films was determined with energy dispersive X-Ray spectroscopy (EDX spectrometer, Rigaku Inc.). The X-ray diffraction experiments used an Empyrean diffractometer (PANalytical) with Cu-K$_a$–radiation. The XRD data were analyzed by High Score Plus and Full Proof programs.

Cross-sections of the multilayered periodic shields were obtained via an OLYMPUS GX41 optical microscope.

The initial magnetization curve, the hysteresis loops, and the static magnetic characteristics including the μ_{max} maximum magnetic permeability, H_c coercive force, and H_s saturation field were measured via the ballistic method. In this aim, samples with an outer diameter of 45 mm and an inner diameter of 25 mm were manufactured. The magnetizing coil was formed via a Ø0.5 mm wire with 100 turns: the measurement coil used Ø0.08 mm wire with of 200 turns amount. The c magnetic characteristics were controlled via the vibration method [19] using a universal cryogenic high-field measurement system (Liquid Helium Free High Field Measurement System by Cryogenic Ltd, London, UK) at room temperature.

Quantitative evaluation of the shielding effectiveness (Ef) used the magnetic field strength or induction ratio in the protected area of space at the H_{ext} (or B_{ext}) shield absence and at the H_{int} (or B_{int}) shield presence [20]:

$$Ef = B_{ext}/B_{int} = H_{ext}/H_{int} \qquad (1)$$

The shield effectiveness research used three mutually perpendicular Helmholtz coils, inducing a three-component permanent magnetic field from 0 to 4500 A/m [20]. The test sample was placed in a uniform magnetic field created by one of the pairs of coils (magnetizing coils) powered by DC B5-86/1 current source and was monitored using a DC M253 ammeter. A switching device changed the direction and type of the current

Eng. Magnetic, Dielectric and Microwave Properties of Ceramics and Alloys   Materials Research Forum LLC
Materials Research Foundations **57** (2019) 57-74   doi: https://doi.org/10.21741/9781644900390-3

(DC, AC) in the magnetizing coils. The second pair of coils (compensation coils) compensates the Earth's magnetic field. The power of the compensation coils used a DC B5-86/1 current source and was controlled with a DC M253 ammeter. The Earth's magnetic field was compensated to at least 5 A/m by adjusting the amount of current and the coil axis direction.

Calculations of shielding effectiveness are based on measurements of the Hall potential in the predetermined central region without a H_{ext} shield and with a H_{int} shield. A Hall sensor was placed in the central zone of the test sample that corresponds to the center located along the axis of the magnetizing coils [21]. The Hall potential measurements used a calibrated Hall element with a sensitivity of 1 mV/20 Oe and V7-34A digital voltmeter or DC V2-39 nanovoltmeter. The Hall sensor was energized with a B5-44A stabilized DC power supply.

The heterogeneity in the distribution of the magnetic field induction along the three axes in both directions is relative to the center of the three coils at 4 cm, 6 cm, 8 cm and 10 cm. It is not more than 0.4%, 0.8%, 1.7% and 3.1%, respectively.

3.   Results and discussions

3.1   Crystal structure and magnetic properties of the $Ni_{1-x}Fe_x$ films

The crystal structure parameters (a, V) were investigated by XRD (PAN analytical in Cu-K_a–radiation) and analyzed by High Score Plus and Full Proof programs. Fig. 1 demonstrates XRD patterns for $Ni_{1-x}Fe_x$ ($0 \leq x \leq 0.5$) films with a thickness of 50 μm. The inserts demonstrate the average chemical composition obtained by EDX (Rigaku Inc.).

The chemical composition of the sample corresponds to the theoretically calculated composition before synthesis. For $Ni_{1-x}Fe_x$ with x = 0.5, the real chemical composition was $Ni_{50.48}Fe_{49.52}$. The deviation from the calculated stoichiometry was less than 0.5 at. %. For $Ni_{1-x}Fe_x$ with x = 0.2, the real chemical composition was $Ni_{79.36}Fe_{20.64}$. The deviation from the calculated stoichiometry was less than 0.7 at. %. The $Ni_{1-x}Fe_x$ with x = 0 consist of pure nickel (100% Ni). For convenience, we will designate these samples as $Ni_{1-x}Fe_x$ x = 0, 0.2 and 0.5 or Ni, $Ni_{80}Fe_{20}$ and $Ni_{50}Fe_{50}$ respectively.

XRD analysis shows that $Ni_{1-x}Fe_x$ ($0 < x < 0.5$) films are based on face-centered cubic lattice (space group Fm-3m). The microstrain (ε) and the average size of the coherent scattering regions ($<D_V>$) characterized the microstructural parameters [22]. All parameters of the crystal structure, relevance factors (R-factors) and fitting coefficients (χ^2) are represented in Table 1.

Eng. Magnetic, Dielectric and Microwave Properties of Ceramics and Alloys   Materials Research Forum LLC
Materials Research Foundations **57** (2019) 57-74   doi: https://doi.org/10.21741/9781644900390-3

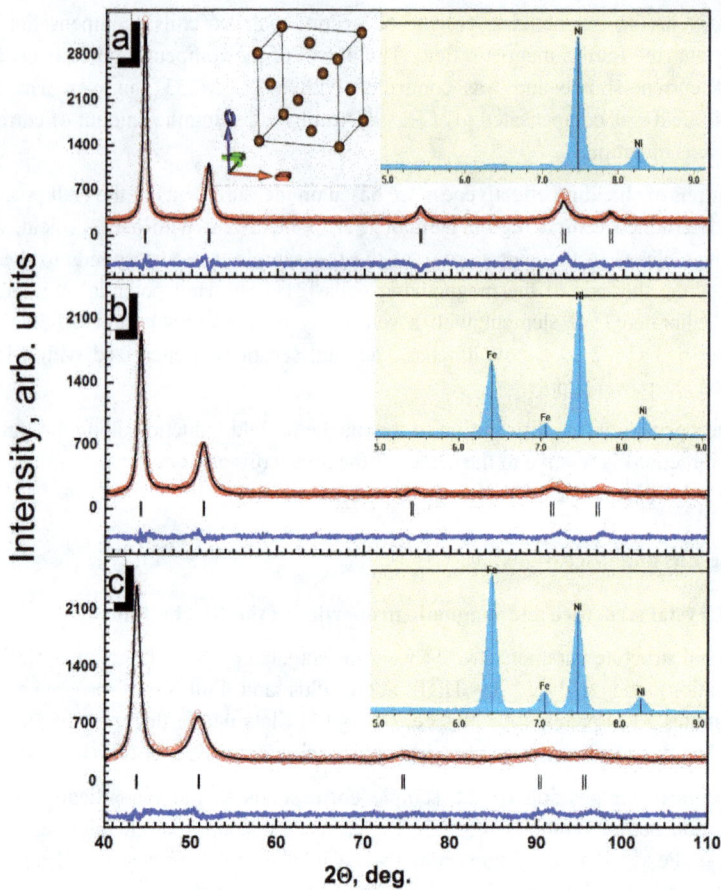

Fig. 1– *The XRD patterns of the Ni (a), $Ni_{80}Fe_{20}$ (b) and $Ni_{50}Fe_{50}$ (c) films (inserts demonstrate EDX spectra of the $Ni_{1-x}Fe_x$ films).*

Eng. Magnetic, Dielectric and Microwave Properties of Ceramics and Alloys   Materials Research Forum LLC
Materials Research Foundations **57** (2019) 57-74   doi: https://doi.org/10.21741/9781644900390-3

Table 1 *Crystal structure parameters, parameters of the integral width $B_{(111)}$ of the (111) line and microstructure and standard factors obtained by Rietveld refinement within Space Group #225 (Fm-3m) at room temperature for Ni 4a (0, 0, 0) and Ni/Fe 4a (0, 0, 0).*

Samples Parameters	Ni	$Ni_{80}Fe_{20}$	$Ni_{50}Fe_{50}$
a, (Å)	3.530(4)	3.573(2)	3.621(3)
V, (Å^3)	43.98(1)	45.62(4)	47.49(7)
$B_{(111)}$, deg.	0.73	1.06	1.42
$<D_V>$, Å	154	83	209
ε, %	0.38	0.58	1.63
B_{iso}	0.315(0)	0.123(0)	0.329(0)
R_p, %	24.9	19.3	18.0
R_{wp}, %	19.9	17.6	16.7
R_{exp}, %	16.05	17.68	14.97
χ^2	1.54	0.989	1.24

The increase in the Fe concentration in $Ni_{1-x}Fe_x$ from x=0 to x=0.5 increases the 'a' parameter of the unit cell from 3.530(4) Å to 3.621(3) Å. This concerns differences in the atomic radius of the Fe (0.126 nm) and Ni (0.124 nm). The unit cell volume also increases with Fe content increase. The micro strain coefficient changes from 0.38 % to 1.63 % as well. However, it is more interesting that the average size of the coherent scattering regions has no correlation with Fe concentration in films (see Table 1). This is difficult to interpret clearly at this stage of research and is probably due to differences in the grain surface defects in the film.

Diffraction peak (111) broadening was used to calculate the microstrain(ε) and the average size of the coherent scattering regions ($<D_V>$). A comparison between LaB6 etalon sample SRM 656b (NIST) was made to separate the ε and $<D_V>$ contribution in the broadening from the diffraction peaks (111) using Williamson – Hall method. Fig. 2 demonstrates the dependency of the diffraction reflection broadening β (θ) as a function of the sin (θ) for the experimental (Ni, $Ni_{80}Fe_{20}$ and $Ni_{50}Fe_{50}$) films and reference sample (LaB_6) at room temperature.

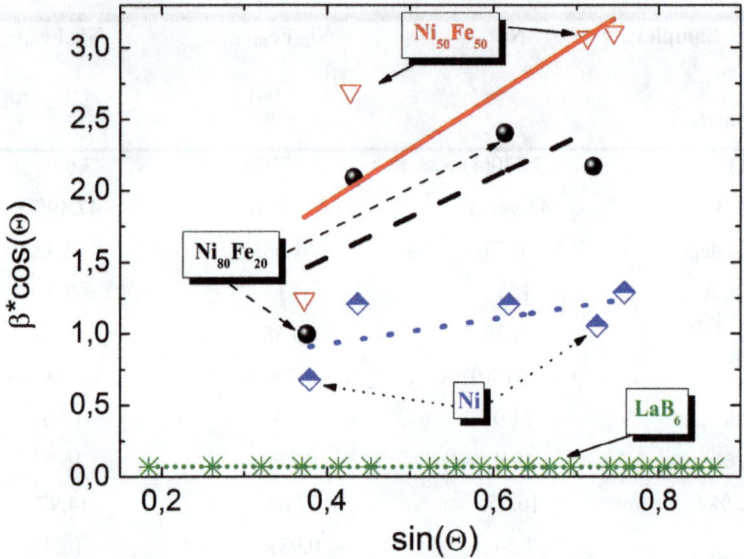

Fig. 2– *Dependence of the diffraction reflections broadening β (θ) as a function of the sin (θ) for the experimental samples and reference sample (LaB₆) at room temperature. Straight lines are the linear approximation.*

Fig. 3 demonstrates concentration dependencies of the main magnetic properties: magnetic permeability (μ), coercivity (H_c) and induction (B) for $Ni_{1-x}Fe_x$ ($0 < x < 0.5$) films with different chemical compositions.

Eng. Magnetic, Dielectric and Microwave Properties of Ceramics and Alloys   Materials Research Forum LLC
Materials Research Foundations **57** (2019) 57-74   doi: https://doi.org/10.21741/9781644900390-3

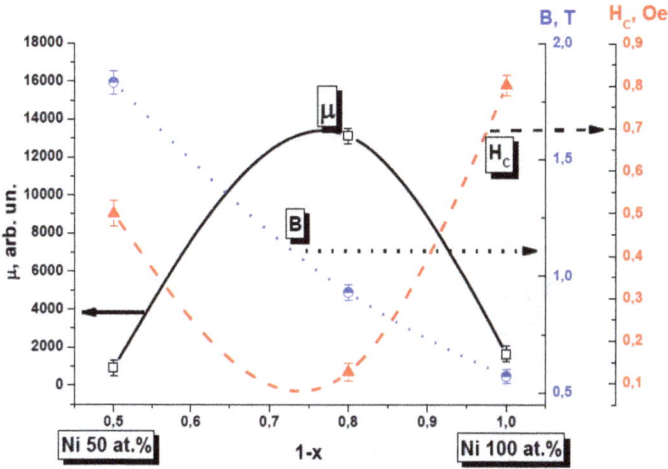

Fig. 3– *Concentration dependences of the main magnetic parameters (μ, B and H_c) for $Ni_{1-x}Fe_x (0 \leq x \leq 0.5)$ films.*

Induction (B) decreases linearly from the 1.83 T to 0.57 T with decreasing iron concentration. This is logical because of the difference between the magnetic moment values of iron and nickel. Magnetic permeability and coercivity are not linear with (1-x). The maximum value of the μ ($>13 \times 10^3$) corresponds to the $Ni_{80}Fe_{20}$ films. This sample has the lowest coercivity value (0.126 Oe). The concentration dependence of the magnetic properties in $Ni_{1-x}Fe_x$ (0<x<0.5) films has complicate behavior and is difficult to interpret unambiguously [23].

3.2   Magnetic properties of the $Ni_{80}Fe_{20}$ films

To interpret the shielding effectiveness, the initial magnetization curve and hysteresis loop characteristics of the $Ni_{80}Fe_{20}$ alloy films were experimentally measured for 100 nm – 80 μm thick samples. The field dependencies of these samples are shown on Fig. 4. The μ_{max} values increased from 350 to 13564 with thickness increasing from 100 nm to 80 μm respectively. Of note, the μ_{max} shifted toward higher fields with increasing thickness. Samples with a low thickness (100 nm and 1 μm) had blurred permeability peak. Thicker samples (10 μm – 80 μm) μ(H) dependencies were Gaussians with a distinct maximum.

Eng. Magnetic, Dielectric and Microwave Properties of Ceramics and Alloys   Materials Research Forum LLC
Materials Research Foundations **57** (2019) 57-74   doi: https://doi.org/10.21741/9781644900390-3

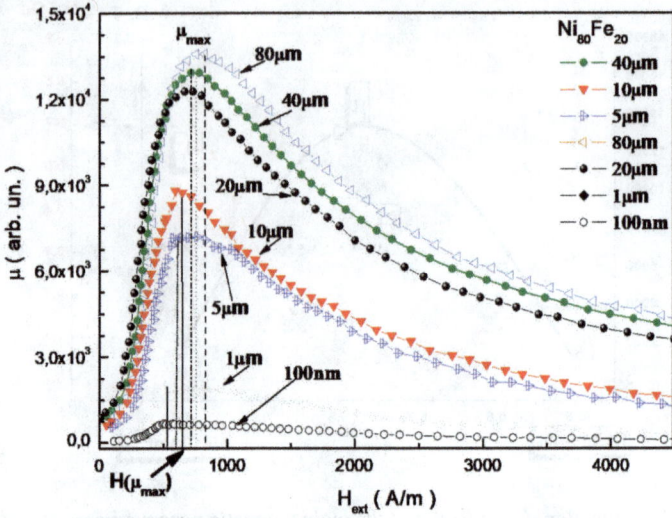

Fig. 4– *Field dependences of magnetic permeability for $Ni_{80}Fe_{20}$ film samples with different thickness (100 nm – 80 μm).*

The thickness dependencies of the coercive force and magnetic permeability for $Ni_{80}Fe_{20}$ film samples are shown in Fig. 5. The abrupt changes in permeability value (μ_{max}) are observed with thickness values of 100 nm-5 μm. Value of μ_{max} increased from 350 (for $Ni_{80}Fe_{20}$ film 100 nm) to 7146 (for $Ni_{80}Fe_{20}$ film 5 μm) when the thickness increased. Thicknesses values of 20-80 μm had the equal permeability values [24]. Table 2 shows the main magnetic parameters (μ_{max}, H_c, and $H_{\mu max}$) for $Ni_{80}Fe_{20}$ films with different thickness values. The coercive force (H_c) sharply decreased (from 1.95 Oe to 0.13 Oe) when the thickness increased from 100 nm to 5 μm. From 5-80 μm, changes in coercive force values were insignificant (0.13-0.126 Oe).

This means that for practical applications (magnetic flux shunting), more perspective (as ferromagnetic layers in multilayered structure) are achieved for ferromagnetic films with partial layer thickness >10-80 μm.

Table 2. *Main magnetic parameters (μ_{max}, H_c, $H_{\mu max}$) for $Ni_{80}Fe_{20}$ films with different thickness.*

$Ni_{80}Fe_{20}$ film thickness, μm	μ_{max}	H_c, Oe	$H_{\mu max}$, A/m
0.1	350	1.95	503
1	1988	1.29	524
5	7146	0.38	574
10	8723	0.13	645
20	12273	0.129	729
40	12919	0.127	779
80	13564	0.126	819

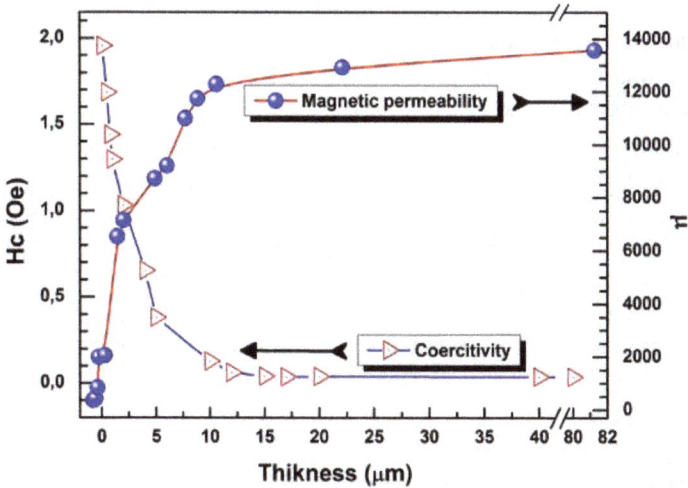

Fig. 5– *Thickness dependences of the coercivity–H_c(left scale) and magnetic permeability –μ (right scale) for $Ni_{80}Fe_{20}$ film samples.*

Eng. Magnetic, Dielectric and Microwave Properties of Ceramics and Alloys   Materials Research Forum LLC
Materials Research Foundations **57** (2019) 57-74   doi: https://doi.org/10.21741/9781644900390-3

3.3   Magnetostatic multilayered shields based on $Ni_{80}Fe_{20}$ films

There is good contrast between the ferromagnetic and copper layers. It can be well distinguished the thickness and the number of ferromagnetic, i.e. thick layers, with highly conductive layers, i.e. thin layers. They are clearly separated in the phase's interfaces. The thicknesses of the partial layers in each sample are equal. In samples with low layer thicknesses with 40 and 80 numbers of layers, the surface relief, roughness, is inherent. The chemical composition of the ferromagnetic layers shows that all samples are $Ni_{80}Fe_{20}$.

A shielding study in multilayered film structures with different numbers and thicknesses of individual ferromagnetic layers showed that the maximum coefficients of the magnetic field shielding, i.e.shielding effectiveness, occur in the samples with the thickest ferromagnetic layers with 80-400 μm. The highest shielding effectiveness of ~ 45.3 was fixed for the sample with maximal thickness and partial ferromagnetic layer, 1x400 μm. The maximum shielding effectiveness for the sample with internal structure 5x80 μm was ~ 31.7. This can be explained in terms of magnetic flux shunting. The thickness of the ferromagnetic shield with high magnetic permeability is the main parameter for the total flux conduction [25]. The minimum value of the shielding effectiveness was detected for the sample with the minimum thickness of the partial ferromagnetic layer, 5 μm, and maximum numbers of the layers, i.e. 80 layers.

It is interesting that the shielding peak, i.e. maximum of the shielding effectiveness is shifted to the higher fields region with an increasing thickness of ferromagnetic layers [26]. This behavior (displacement of the local maxima) in the external magnetic fields can be explained by the non-linear distribution of the magnetic permeability by the shield thickness [1] and non-equilibrium transition of the shield material in the magnetic saturation state [27]. The main shielding parameters (Eff$_{max}$, H$_{Eff.max}$) for $Ni_{80}Fe_{20}$/Cu multi-layered structures with different numbers and thickness of the ferromagnetic layer (total thickness of ferromagnetic layers is 400 μm) are presented in Table 3.

There is decreased shielding effectiveness for magnetic fields above critical values, i.e. field strength at which the sample shows maximum effectiveness [28-33]. This may be due to a rapid decrease in the magnetic permeability at higher fields ($H \gg H_{\mu max}$).

Eng. Magnetic, Dielectric and Microwave Properties of Ceramics and Alloys   Materials Research Forum LLC
Materials Research Foundations **57** (2019) 57-74   doi: https://doi.org/10.21741/9781644900390-3

Table 3. *Main shielding parameters for $Ni_{80}Fe_{20}/Cu$ multi-layered structures.*

$Ni_{80}Fe_{20}/Cu$ multi-layered structures	Eff_{max}	$H_{Eff.max}$, A/m
1x400 µm$Ni_{80}Fe_{20}$	45.3	1420
5x80 µm $Ni_{80}Fe_{20}$	40.6	1240
10x40 µm $Ni_{80}Fe_{20}$	34.8	711
40x10 µm $Ni_{80}Fe_{20}$	28.6	531
80x5 µm $Ni_{80}Fe_{20}$	27.4	531

Conclusion

The $Ni_{1-x}Fe_x$ (x=0, 0.2 and 0.5) films were produced via electrodeposition. The chemical composition of the samples corresponds to the theoretically calculated composition before synthesis. For $Ni_{1-x}Fe_x$ (x=0.5), the real chemical composition was $Ni_{50.48}Fe_{49.52}$ (deviation from the calculated stoichiometry is less than 0.5 at. %). For $Ni_{1-x}Fe_x$ (x=0.2), the real chemical composition was $Ni_{79.36}Fe_{20.64}$ (deviation from the calculated stoichiometry is less than 0.7 at. %). The $Ni_{1-x}Fe_x$(x=0) totally consist of pure nickel (100% Ni). The $Ni_{1-x}Fe_x$ (0≤x≤0.5) films were based on a face-centred cubic lattice (space group Fm-3m).The increase of the Fe concentration in $Ni_{1-x}Fe_x$ from x=0 to x=0.5 increases the '*a*' parameter of the unit cell from 3.530(4) Å to 3.621(3) Å respectively. This shows a difference in the atomic radius of the Fe (0.126 nm) and Ni (0.124 nm). The unit cell volume increases with increasing Fe content. The microstrain (ε) and the average size of coherent scattering regions ($<D_V>$) were studied. Broadening of the diffraction peaks (111) were used to calculate of the microstrain (ε) and the average size of the coherent scattering regions ($<D_V>$). Microstrain coefficient changes from 0.38 % to 1.63 %. The average size of the coherent scattering regions has no correlations with Fe concentration in the films.

The main magnetic parameters (magnetic permeability, coercivity and induction) were studied as a function of the Fe-concentration. The B value decreases linearly from 1.83 T to 0.57 T as iron concentration decreases. This is logical and is due to the differences between the magnetic moment values of iron and nickel. The magnetic permeability and coercivity are not linear functions of the Fe-concentration. The maximum value of the μ ($>13 \times 10^3$) corresponds to $Ni_{80}Fe_{20}$ films. This sample has a minimum value of the coercivity (0.126 Oe). In practical applications (as magnetic shields) films (and multilayered structures) based on $Ni_{80}Fe_{20}$ electrodeposited layers aremore appropriate. Magnetic permeability and coercivity were measured as a function of the thickness of the $Ni_{80}Fe_{20}$ films. The field dependences of the initial magnetic permeability for the $Ni_{80}Fe_{20}$ films with different thickness from 100 nm up to 80 µm were investigated. The μ_{max} values increase from 350 to 13564 as the thickness increases from 100 nm to 80 µm. The μ_{max} shifts toward higher fields as the thickness increases. For thin samples (100 nm and 1 µm), the permeability peak is blurred. Thicker sample (10 µm – 80 µm), µ(H) dependencies have Gaussians shapes with a distinct maximum. This means that for practical applications (for magnetic flux shunting), more perspective is achieved with more ferromagnetic layers in the multilayered structure partial layer thickness >10-80 µm.

Multilayered shields based on $Ni_{80}Fe_{20}$/Cu films were formed for shielding measurements. The shielding effectiveness of the $Ni_{80}Fe_{20}$ / Cu multilayered samples with the same total thickness of the ferromagnetic layers (400 µm) and with different numbers of the partial ferromagnetic layers (1-80) and their thickness (400-5 µm) was investigated. The highest shielding effectiveness (~ 45.3) was fixed for the sample with the maximal thickness partial ferromagnetic layer (1x400 µm). The maximum shielding effectiveness for the sample with an internal structure of 5x80 µm was ~ 31.7. This can be explained in terms of the theory of magnetic flux shunting (thickness of the ferromagnetic shield with high magnetic permeability is a main parameter for the total flux conduction). The minimum value of the shielding effectiveness was detected for the sample with the minimum thickness of the partial ferromagnetic layer (5 µm) and maximum numbers of the layers (80 layers). It is interesting that the shielding peak (maximum of the shielding effectiveness) is shifted to higher field regions as the layer thickness increases. This behavior (displacement of the local maxima) in the external magnetic fields can be explained by the non-linear distribution of the magnetic permeability by the shield thickness and non-equilibrium transition of the shield material in the magnetic saturation state. There was decreased shielding effectiveness for magnetic fields above critical values (field strength at which the sample shows maximum effectiveness). That may be due to rapid decreasing of the magnetic permeability at higher fields ($H \gg H_{\mu max}$).

Acknowledgement

This work was carried out with financial support in part from the Ministry of Science and Higher Education of the Russian Federation (task in SUSU 4.1346.2017/4.6). The work was supported by the Russian Foundation for Basic Research (project No. 19-32-50029, program "mol_nr").

Eng. Magnetic, Dielectric and Microwave Properties of Ceramics and Alloys   Materials Research Forum LLC
Materials Research Foundations **57** (2019) 57-74   doi: https://doi.org/10.21741/9781644900390-3

References

[1] Y. Jiraskova,J. Bursik,I. Turek,M. Hapla,A. Titov,O. Zivotsky,Phase   and   magnetic
studies of the high-energy alloyed Ni–Fe, J. Alloys Compd. 594 (2014) 133.
https://doi.org/10.1016/j.jallcom.2014.01.138

[2] X. Zhao,Y. Dang,H. Yin,Y. Yuan,J. Lu,Z. Yang,Y. Gu,Evolution   of   the
microstructure and microhardness of a new wrought Ni–Fe based superalloy during
high   temperature   aging,   J.   Alloys   Compd.   644   (2015)   66.
https://doi.org/10.1016/j.jallcom.2015.04.184

[3] V. Torabinejad,A.S. Rouhaghdam,
M. Aliofkhazraei,M.H. Allahyarzadeh,Electrodeposition of Ni–Fe and Ni–Fe-(nano
Al$_2$O$_3$)   multilayer   coatings,   J.   Alloys   Compd.   657   (2016)   526.
https://doi.org/10.1016/j.jallcom.2015.10.154

[4] L.K. Béland,G.D. Samolyuk,R.E. Stolle,Differences in the accumulation of ion-beam
damage in Ni and NiFe explained by atomistic simulations, J. Alloys Compd. 662
(2016) 415. https://doi.org/10.1016/j.jallcom.2015.11.185

[5] V. Torabinejad,M. Aliofkhazraei,S. Assareh,M.H. Allahyarzadeh,A.S. Rouhaghdam,
Electrodeposition of Ni-Fe alloys, composites, and nano coatings–A review, J. Alloys
Compd. 691 (2017) 841. https://doi.org/10.1016/j.jallcom.2016.08.329

[6] S.S. Grabchikov, A.V. Trukhanov, S.V. Trukhanov, I.S. Kazakevich, A.A. Solobay,
V.T. Erofeenko, N.V. Vasilenkov,Effectiveness of the magnetostatic shielding by the
cylindrical   shells,   J.   Magn.   Magn.   Mater.   398   (2016)   49.
https://doi.org/10.1016/j.jmmm.2015.08.122

[7] M.H. Al-Saleh,Electrical and electromagnetic interference shielding characteristics of
GNP/UHMWPE   composites,J.   Phys.   D: Appl.   Phys.49   (2016)   19.
https://doi.org/10.1088/0022-3727/49/19/195302

[8] T.J. Sumner, J.M. Pendlebury,K.F. Smith,Convectional magnetic shielding,J. Phys. D:
Appl. Phys.20 (1987) 1095. https://doi.org/10.1088/0022-3727/20/9/001

[9] J. Füzi, A. Iványi, Zs. Szabó,Magnetic force computation with hysteresis, J. Magn.
Magn. Mater. 254-255 (2003) 237. https://doi.org/10.1016/S0304-8853(02)00777-1

[10]   D. Bavastro, A. Canova, L. Giaccone,M. Manca,Numerical and experimental
development of multilayer magnetic shields,Electric Power Systems Research 116
(2014) 374. https://doi.org/10.1016/j.epsr.2014.07.004

[11]   L. Hasselgren, J. Luomi,Geometrical aspects of magnetic shielding at extremely
low   frequencies,   IEEE   Trans.   Electromagn.   Compat.   37   (1995)   409.
https://doi.org/10.1109/15.406530

[12] O. Bottauscio,   M. Chiampi,   D. Chiarabaglio,   F. Fiorillo,   L. Rocchino, M. Zucca,Role of magnetic materials in power frequency shielding: numerical analysis and experiments, IEE Proceedings of Generation, Transmission and Distribution, 148 (2001) 104. https://doi.org/10.1049/ip-gtd:20010162

[13] L. Sandrolini,U. Reggiani, A. Ogunsola,Modelling the electrical properties of concrete for shielding effectiveness prediction,J. Phys. D: Appl. Phys.,40 (2007) 17. https://doi.org/10.1088/0022-3727/40/17/053

[14] A.V. Trukhanov,   S.S. Grabchikov,   A.A. Solobai,   D.I. Tishkevich, S.V. Trukhanov, E.L. Trukhanova,AC and DC-shielding properties for the $Ni_{80}Fe_{20}$/Cu film structures,J. Magn. Magn. Mater. 443 (2017) 142. https://doi.org/10.1016/j.jmmm.2017.07.053

[15] S.V. Trukhanov,   N.V. Kasper,   I.O. Troyanchuk,   M. Tovar,   H. Szymczak, K. Bärner,Evolution of magnetic state in the $La_{1-x}Ca_xMnO_{3-\gamma}$ (x=0.30, 0.50) manganites depending on the oxygen content,J. Sol. State Chem. 169 (2002) 85. https://doi.org/10.1016/S0022-4596(02)00028-2

[16] V.D. Doroshev, V.A. Borodin, V.I. Kamenev, A.S. Mazur, T.N. Tarasenko, A.I. Tovstolytkin,S.V. Trukhanov,Self-doped lanthanum manganites as a phase-separated system: Transformation of magnetic, resonance, and transport properties with doping and hydrostatic compression,J. Appl. Phys. 104 (2008) 093909. https://doi.org/10.1063/1.3007993

[17] S.V. Trukhanov, I.O. Troyanchuk, I.M. Fita, H. Szymczak, K. Bärner,Comparative study of the magnetic and electrical properties of $Pr_{1-x}Ba_xMnO_{3-\delta}$ manganites depending on the preparation conditions,J. Magn. Magn. Mater. 237 (2001) 276. https://doi.org/10.1016/S0304-8853(01)00477-2

[18] A.V. Trukhanov, S.S. Grabchikov, A.N. Vasiliev, S.A. Sharko, N.I. Mukhurov, I.V. Gasenkova,Specific features of formation and growth mechanism of multilayered quasi-one-dimensional (Co-Ni-Fe)/Cu systems in pores of anodic alumina   matrices,Crystallogr.   Reports   59   (2014)   744. https://doi.org/10.1134/S1063774514050216

[19] S.V. Trukhanov,   A.V. Trukhanov,   A.N. Vasiliev,   H. Szymczak,Frustrated exchange interactions formation at low temperatures and high hydrostatic pressures in $La_{0.70}Sr_{0.30}MnO_{2.85}$,JETP   111   (2010)   209. https://doi.org/10.1134/S106377611008008X

[20] E.M. Purcell, Electricity and Magnetism, McGraw Hill, (1985).

[21] S.V. Trukhanov,   L.S. Lobanovski,   M.V. Bushinsky,   I.O. Troyanchuk, H. Szymczak,Magnetic phase transitions in the anion-deficient $La_{1-x}Ba_xMnO_{3-x/2}$

($0 \leq x \leq 0.50$) manganites,J. Phys.: Condens. Matter, 15 (2003) 1783. https://doi.org/10.1088/0953-8984/15/10/324

[22] S.V. Trukhanov, A.V. Trukhanov, H. Szymczak, C. E. Botez, A. Adair,Magnetotransport properties and mechanism of the A-site ordering in the Nd–Ba optimal-doped manganites,J. Low Temp. Phys. 149 (2007) 185. https://doi.org/10.1007/s10909-007-9507-6

[23] S.V. Trukhanov, L.S. Lobanovski, M.V. Bushinsky, V.A. Khomchenko, N.V. Pushkarev, I.O. Tyoyanchuk, A. Maignan, D. Flahaut, H. Szymczak, R. Szymczak,Influence of oxygen vacancies on the magnetic and electrical properties of $La_{1-x}Sr_xMnO_{3-x/2}$manganites,Eur. Phys. J. B 42 (2004) 51. https://doi.org/10.1140/epjb/e2004-00357-8

[24] S.V. Trukhanov, I.O. Troyanchuk, A.V. Trukhanov, I.M. Fita, A.N. Vasil'ev, A. Maignan, H. Szymczak,Magnetic properties of $La_{0.70}Sr_{0.30}MnO_{2.85}$anion-deficient manganite under hydrostatic pressure,JETP Lett. 83 (2006) 33. https://doi.org/10.1134/S0021364006010085

[25] A.O. Shiryaev, K.N. Rozanov, S.A. Vyzulin, A.L. Kevraletin, N.E. Syr'ev, E.S. Vyzulin, E. Lahderanta, S.A. Maklakov, A.B. Granovsky,Magnetic resonances and microwave permeability in thin Fe films on flexible polymer substrates,J. Magn. Magn. Mater. 461 (2018) 76. https://doi.org/10.1016/j.jmmm.2018.04.059

[26] L.D. Geng, Y.M. Jin,Controlling 180° transverse domain wall structure,J.Magn.Magn. Mater. 468 (2018) 246. https://doi.org/10.1016/j.jmmm.2018.08.021

[27] Y. Wang, Y. Wen, P. Li,Analytical and experimental study of the improved magnetic field sensitivity for nanocrystalline soft magnetic alloy and coil laminate with different layers,J.Magn.Magn. Mater. 474 (2019) 36. https://doi.org/10.1016/j.jmmm.2018.10.144

[28] S.V. Trukhanov, A.V. Trukhanov, V.A. Turchenko, An.V. Trukhanov, D.I. Tishkevich, E.L. Trukhanova, T.I. Zubar, D.V. Karpinsky, V.G. Kostishyn, L.V. Panina, D.A. Vinnik, S.A. Gudkova, E.A. Trofimov, P. Thakur, A. Thakur, Y. Yang, Magnetic and dipole moments in indium doped barium hexaferrites, J. Magn. Magn. Mater. 457 (2018) 83. https://doi.org/10.1016/j.jmmm.2018.02.078

[29] V.A. Turchenko, A.V. Trukhanov, I.A. Bobrikov, S.V. Trukhanov, A.M. Balagurov, Investigation of the crystal and magnetic structures of $BaFe_{12-x}Al_xO_{19}$ solid solutions (x = 0.1-1.2),Crystallogr. Rep.60 (2015) 629. https://doi.org/10.1134/S1063774515030220

[30] S.V. Trukhanov,   A.V. Trukhanov,   V.G. Kostishyn,   L.V. Panina,
An.V. Trukhanov,V.A. Turchenko,   D.I. Tishkevich,   E.L. Trukhanova,
O.S. Yakovenko, L.Yu. Matzui,D.A. Vinnik, D.V. Karpinsky, Effect of gallium
doping on electromagnetic properties of barium hexaferrite, J. Phys. Chem. Sol.111
(2017) 142. https://doi.org/10.1016/j.jpcs.2017.07.014

[31] V.A. Turchenko,   A.V. Trukhanov,   I.A. Bobrikov,   S.V. Trukhanov,
A.M. Balagurov, Study of the crystalline and magnetic structures of $BaFe_{11.4}Al_{0.6}O_{19}$
in   a   wide   temperature   range,J.   Surf.   Investig.   9   (2015)   17.
https://doi.org/10.1134/S1027451015010176

[32] A.V. Trukhanov, V.G. Kostishyn, L.V. Panina, S.H. Jabarov, V.V. Korovushkin,
S.V. Trukhanov, E.L. Trukhanova, Magnetic properties and Mössbauer study of
gallium doped M-type barium hexaferrites, Ceram. Int. 43 (2017) 12822.
https://doi.org/10.1016/j.ceramint.2017.06.172

[33] I.O. Troyanchuk, S.V. Trukhanov, D.D. Khalyavin, H. Szymczak, Magnetic
properties of anion deficit manganites $Ln_{0.55}Ba_{0.45}MnO_{3-\gamma}$ (Ln=La, Nd, Sm, Gd,
$\gamma \leqslant 0.37$), J. Magn. Magn. Mater. 208 (2000) 217. https://doi.org/10.1016/S0304-
8853(99)00529-6

Eng. Magnetic, Dielectric and Microwave Properties of Ceramics and Alloys   Materials Research Forum LLC
Materials Research Foundations **57** (2019) 75-88   doi: https://doi.org/10.21741/9781644900390-4

Chapter 4

The Development of Capacitor Materials Technology

Maciej Jaroszewski[1,a], Jan Ziaja[1,b], Przemysław Ranachowski[2,c], Charanjeet Singh[3,d]

[1]Wroclaw University of Science and Technology, Faculty of Electrical Engineering, 27 Wybrzeze Wyspianskiego Str., 50-370 Wroclaw, Poland

[2]Polish Academy of Sciences, Institute of Fundamental Technological Research, 5b Pawińskiego Str., 02-106 Warsaw, Poland

[3]Department of Electronics and Communication Engineering, Lovely Professional University, Phagwara Punjab, India

[a]maciej.jaroszewski@pwr.edu.pl, [b]jan.ziaja@pwr.edu.pl, [c]pranach@ippt.pan.pl, [d]rcharanjeet@gmail.com

Abstract

We are in the era of energy storage devices and lot of research is going on cpacitors to bridge the gap for rebuttals found in conventional capacitors. Herein, basic properties of capacitors are mentioned. The performance parameters of barium titanate are presented which is used as dielectric material in capacitors. The manufacturing process of multi-layer ceramic capacitors is discussed. The detailed description of supercapacitor behavior is mentioned with necessary mechanisms. The review of different materials incorporated in electrodes of the supercapacitor has been presented.

Keywords

Capacitor Basics, Ceramic Capacitor, Supercapacitor, Electrode Materials in Supercapacitor

Contents

Eng. Magnetic, Dielectric and Microwave Properties of Ceramics and Alloys   Materials Research Forum LLC
Materials Research Foundations **57** (2019) 75-88   doi: https://doi.org/10.21741/9781644900390-4

1.   A dielectric material in an electric field

In the general definition, dielectrics are materials in which - under the influence of the electric field - the carriers of electric charges cannot move on a macroscopic scale. In these materials, however, the external electric field causes changes in the electric charge on a microscopic scale. In non-polar dielectrics, the electric field causes atoms to deform by moving the centers of atoms along the field lines. The result is an electrical dipole with an induced dipole moment

$$\vec{p} = q \cdot \vec{l} \tag{1}$$

In addition to the deformation of electron shells, the second effect of the electric field on the dielectric may be the deformation of bonds in the molecule. In this case, an additional dipole moment occurs, related to the movement of atoms from their equilibrium position. The total dipole moment of the molecule will be the sum of both dipole moments, and this phenomenon is called the deformation polarization. In polar dielectrics - in which molecules have a permanent dipole moment - an external electric field causes the arrangement of dipoles, causing an increase in the dipole moment of the entire medium. This phenomenon is called orientation (dipole) polarization.

In macroscopic terms, the phenomenon of polarization is described by the polarization vector P. This vector is defined as the dipole moment of the unit of the volume of the substance in which there is N elemental dipoles p_i, i.e:

$$\vec{P} = \lim_{\Delta V \to 0} \frac{1}{\Delta V} \sum_{i=0}^{N} \vec{p_i} \tag{2}$$

In a polarized dielectric, dipoles form a bound charge on its surface that is the source of its own electric field (polarization) E_p. This field is superimposed on the external force field E, the source of which is free charges, causing the dielectric effect of the electric field E_i to affect the interior of the dielectric

$$\vec{E}_i = \vec{E} + \vec{E}_p \tag{3}$$

The polarization vector (associated) is associated with the polarization vector P. For the isotropic dielectrics, the polarization vector is proportional to the intensity of the electric field inside the dielectric, i.e.

$$P = \varepsilon_0 \chi_E \tag{4}$$

Where ε_0 is the electric permmitivity of the vacuum and χ is the dielectric susceptibility. The normal component of the polarization vector equals the density of the polarization charge accumulated on the surface of the dielectric. The charge accumulated on the electrodes, being the source of the external field, is assigned the vector $\varepsilon_0 \vec{E}$. These two vectors associate the dielectric shift vector D with which the normal component corresponds to the density of the total charge

$$\vec{D} = \varepsilon_0 \vec{E} + \vec{P} = (1 + \chi) \cdot \varepsilon_0 \vec{E} = \varepsilon \cdot \varepsilon_0 \vec{E} \tag{5}$$

2.   Capacitors: basic properties / dependencies

The occurrence of the polarization phenomenon in the dielectric inserted in between the electrodes with different potentials results in the creation of an additional (in relation to the electric field generated by electrodes in a vacuum) fields from polarization charges, the opposite direction to the field of excitation. The field inside the dielectric is corrected by a factor of $1/\varepsilon$. The presence of dielectric results in a decrease in the potential difference between the capacitor electrodes, but due to the constancy of the voltage supplying the system, the capacity of the capacitor with the C_d dielectric will change as compared to the vacuum (air) capacitor C_0:

$$C_d = \varepsilon C_0 \tag{6}$$

The capacitance of the capacitor is therefore dependent on the dielectric permittivity, but also on the surface of the electrodes S and the distance between the electrodes d. For a flat capacitor, the capacitance can be determined from the equation

$$C = C_d = \varepsilon \varepsilon_0 \frac{S}{d} \tag{7}$$

The energy W stored in such a capacitor depends on its electrical capacity C and the voltage between its electrodes V:

$$W=0.5C{\cdot}V^2 \tag{8}$$

Inserting into the above equation the formula for the capacitance of a flat capacitor and taking into account that the voltage is equal to the product of field intensity and the distance between electrodes, and that the dielectric volume in the capacitor V is equal to the product of the electrode surface and the distance (d) between them - we get a dependence on the energy stored in the capacitor as

$$W=0.5(E{\cdot}d)^2V \tag{9}$$

It results from the above equation that the energy density of the electric field (w) that can accumulate in the capacitor is

$$w=0.5(E{\cdot}d)^2 \tag{10}$$

3.   Ceramic capacitors

The wide range of ceramic compositions and related diverse dielectric behavior has contributed to the use of ceramic capacitors in many extreme conditions, such as high temperatures or high voltages [1, 2].

Modern ceramic materials used to build capacitors are characterized by electrical permittivity from several to several tens of thousands, with its temperature variability of several ppm to one percent. These materials are divided into two main classes depending on the value of electrical permittivity and the dielectric loss factor.

Class I ceramics are characterized by low values of electric permittivity and dielectric loss factor. This group includes early ceramic materials such as porcelain or steatites and present ceramics based on oxides such as TiO_2 or $CaTiO_3$.

Class II ceramics are materials with very high values of electric permittivity amounting to several - tens of thousands and loss rates of hundredth parts. These materials are based on the $BaTiO_3$ bar titanate. This compound has ferroelectric properties, which is why it has very high values of electrical permittivity. Permittivity is related to its crystal structure, which undergoes a transformation during cooling. This entails rapid changes in electrical permittivity, as shown in Table 1.

Table 1. Barium titanate with dielectric constant as a function of temperature.

Temperature Change (Approx.)	Change in Dielectric Constant (Approx.)	Crystal Structure
-160 to -93	1000 to2700	Rhombohedral
-93 to 0	2700 to 4700	Orthorhombic
0 to 120	4700 to9750	Tetragonal
120 to 160	9750 to3000	cubic

To reduce the unwanted strong temperature dependence of electrical permittivity, barium titanate compositions are formed with other perovskites or by adding admixtures of other elements. The effect is the shift and damping of the dielectric peaks.

4.   Ceramic capacitors production technologies

4.1   Disk capacitors

Ceramic capacitors are characterized by relatively simple and inexpensive production technology. The technology for the production of monolithic (disc) capacitors is similar to the conventional process for the production of ceramic materials. Dielectric powders being input raw materials of ceramics are purchased from commercial suppliers or obtained on site by means of various chemical processes and selected (grouped) in accordance with the manufacturer's reserved formula. They are required to show up:

- high chemical purity,

- sufficiently small size of particles (grains) of identical size and shape,

- low agglomeration capacity of powder particles,

- repeatability of parameters,

- the lowest possible production / purchase costs.

In the first stage of the ceramic production process, the input raw materials are milled in order to de-agglomerate them and ensure a homogeneous and narrow grain size distribution. After mixing, the raw materials are liquefied with dispersing agents and

Eng. Magnetic, Dielectric and Microwave Properties of Ceramics and Alloys   Materials Research Forum LLC
Materials Research Foundations **57** (2019) 75-88   doi: https://doi.org/10.21741/9781644900390-4

liquids. Then, from the slurry - in the spray dryer - granules are obtained which are subject to forming. Raw green-shaped fittings are obtained by uniaxial pressing. The compacts are then fired in accordance with the appropriate firing curve. Metallic contact layers are applied on the end faces. Passivation layers are applied on external surfaces, which protect against environmental, electrical and thermo-mechanical stresses.

4.2   Multi-layer ceramic capacitors

The limitation of obtaining large values of capacitance in disk capacitors results directly from the relation (7). In order to increase the capacity, it would be necessary to increase the surface of the electrodes or the dielectric permittivity of the ceramic, or to reduce its thickness. The solution is the production of multi-layer ceramic capacitors (MLCC). Their idea lies in the parallel connection of many thin-film ceramic capacitors, as shown in Figure 1. The capacitance can be expressed by the following expression:

$$\text{Capacitance} = \frac{\text{Number of Layers x Dielectric Constant x Active Area}}{\text{Dielectric Thickness}} \qquad (11)$$

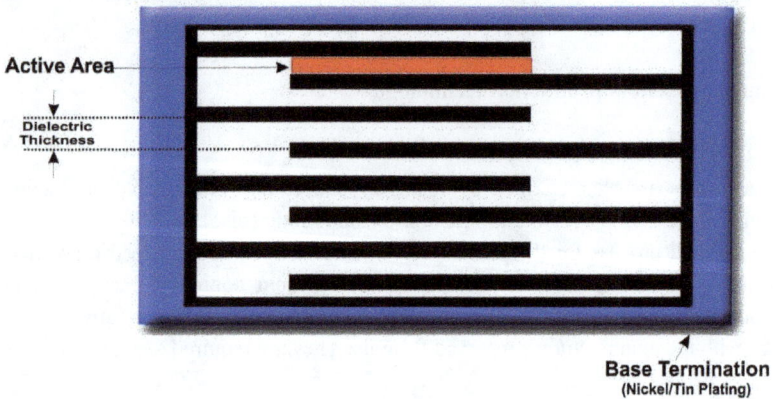

Figure 1. MLCC capacitor.

The technology of their production is similar to the technology of disk capacitors, especially in the first stages of obtaining ceramic material: input raw materials, their grinding and liquefaction. The next stages of the technological process are characterized by a much greater degree of difficulty compared to the technology of monolithic ceramics. In particular, it is difficult technologically to obtain a ceramic tape, apply it to the electrodes by screen printing, precise lamination of individual layers, and later their combined firing. The significant difference between the firing temperature of the ceramic and the melting temperature of the metal used for the electrodes should be particularly emphasized here. In addition, phenomena related to the diffusion of the electrode material into the dielectric ceramic are also added. The technological process is shown in Figure 2.

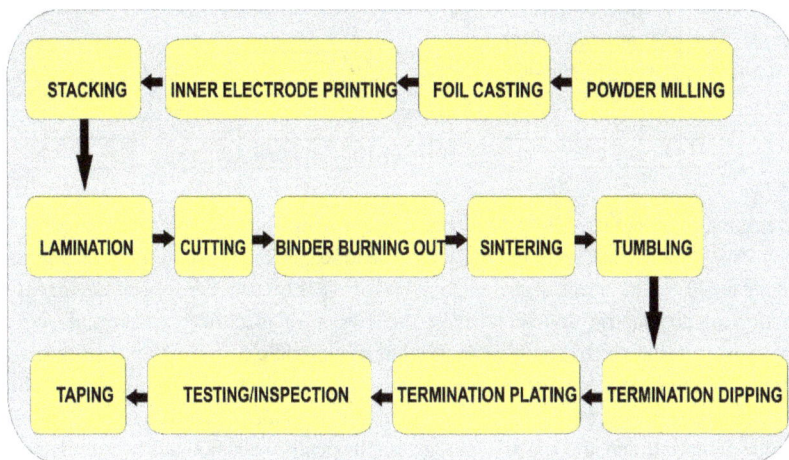

Figure 2. MLCC manufacturing process.

5.   Supercapacitors

The supercapacitor (also called the ultracapacitor) is a specific electrolytic capacitor with an extremely large electrical capacity (up to several thousand farads), high energy density and very high power density. It connects in a sense the features of the battery and the traditional capacitor. High energy density indicates the ability to store energy, while the power density is related to the way the source is used - high power density indicates the possibility of high energy consumption in a short time (and therefore the possibility of charging and discharging with large currents, i.e. getting a quick load exchange). In comparison to batteries, their main advantage, from the point of view of using them as energy storage, is the very high current) with which they can work. Due to this characteristic feature, supercapacitors fill the void between energy storage devices between batteries and traditional capacitors (Table 2).

Eng. Magnetic, Dielectric and Microwave Properties of Ceramics and Alloys   Materials Research Forum LLC
Materials Research Foundations **57** (2019) 75-88   doi: https://doi.org/10.21741/9781644900390-4

Table 2. Energy density as a function of power density for different devices.

Sr. No.	Device	Energy Density $(Wh.Kg^{-1})$ (Approx.)	Power Density $(W.Kg^{-1})$ (Approx.)
1.	Conventional/ ceramic capacitor	10^{-2} to $10^{-1.28}$	$10^{3.5}$ to 10^{7}
2.	Electrochemical Capacitor	$10^{-1.25}$ to $10^{1.125}$	$10^{2.75}$ to $10^{6.25}$
3.	Battery	$10^{0.80}$ to $10^{2.25}$	$10^{0.63}$ to $10^{2.63}$
4.	Fuel Cell	$10^{2.19}$ to $10^{3.08}$	$10^{0.4}$ to $10^{2.3}$

Conventionally, supercapacitors have been categorized into two types: electrochemical capacitors with a double layer (EDLC) and pseudocapacitors based on the charge redox mechanism [3]. The electrolytic capacitors of EDLC use the phenomenon of the formation of an electric double layer at the border of electrode - electrolyte centers, which consist in the electrostatic accumulation of electric charges within it (Figure 3). In this type of supercapacitors, the electrode material is not electrochemically active. During the supercapacitor operation, therefore, there are no electrochemical reactions in the electrode material, and the electric charge is physically accumulated at the electrode / electrolyte interface. In charging and discharging processes, there is no flow (exchange) of charge between the electrolyte and the electrodes (also in the electrode / electrolyte interphase). In contrast, in pseudocapacitors, redox reactions take place on the electrode material. The result is the flow of Faradaic current through the double layer and through the electrolyte.

The capacity of the capacitor depends directly on the surface of the electrodes, and inversely proportional to the distance. Thus the electrodes are made of materials with a significantly developed active surface. Using advanced nanotechnologies, electrodes are produced with gigantic specific surfaces (exceeding even 2,000 m^2 per one gram of electrode), and thus also huge capacities. Organic electrolytes are used as electrolytes (when using higher values of operating voltages - 2.7 to -2.8 V - thanks to which higher energy densities are achieved) or water electrolytes (the operating voltage is limited to 0, 7 ÷ 0.8 V to avoid electrolysis). Schematic diagram of the principle of supercapacitor operation is shown in Figure 4.

Eng. Magnetic, Dielectric and Microwave Properties of Ceramics and Alloys   Materials Research Forum LLC
Materials Research Foundations **57** (2019) 75-88   doi: https://doi.org/10.21741/9781644900390-4

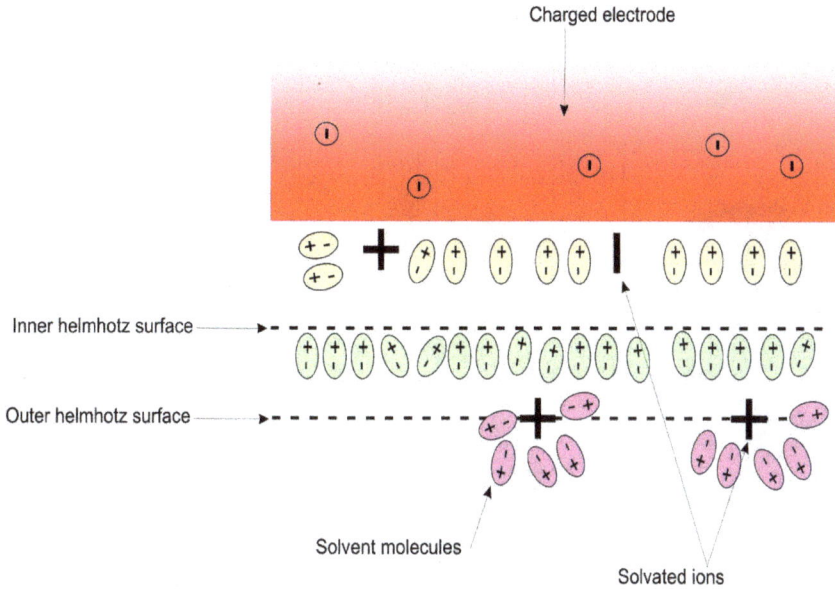

Figure 3. Diagram of the double layer at the electrode-electrolyte interface.

5.1   Material of supercapacitors electrodes

Supercapacitor electrodes must have a large active surface and porosity to ensure the possibility of storage and separation, at the electrode-electrolyte interphase, a correspondingly large value of the electric charge. However, in practice, not all of the specific surface area is electrochemically available for the electrolyte. In order to obtain the largest possible electrochemical active surface, the pore size of the electrode material must be as close as possible to the size of the electrolyte ions. Hence the necessity of using nanotechnology for the production of electrodes. The development of material technology makes it possible to obtain supercapacitors with increasing capacity, also dependent on the base material of electrodes, as shown in Table 3.

Eng. Magnetic, Dielectric and Microwave Properties of Ceramics and Alloys   Materials Research Forum LLC
Materials Research Foundations **57** (2019) 75-88   doi: https://doi.org/10.21741/9781644900390-4

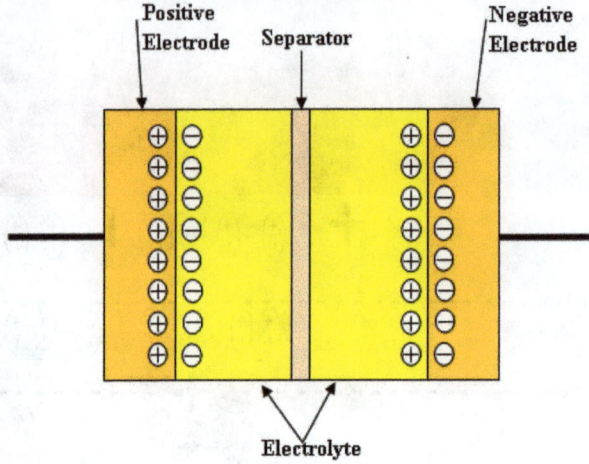

(a) Charging

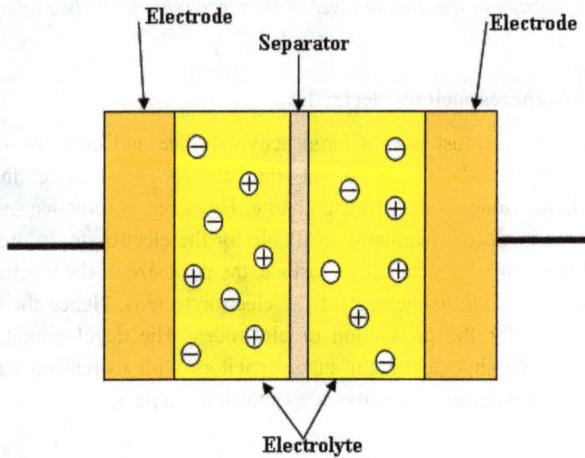

(b) Discharging

Figure 5. Supercapacitor operation.

Eng. Magnetic, Dielectric and Microwave Properties of Ceramics and Alloys   Materials Research Forum LLC
Materials Research Foundations **57** (2019) 75-88   doi: https://doi.org/10.21741/9781644900390-4

Table 3. Specific capacitance and current for different types of electrode materials.

Sr. No.	Material	Specific Capacitance (F/g)	Specific Reference Current (A/g)
1	PANI-IL [4]	662	1
2	Carbon/PANI [5]	747	0.1
3	RGO/PANI [6]	1182	1
4	PANI-NFS/GF [7]	1474	0.5
5	PMTA@CNT/RGO [8]	658	0.5
6	S-g-A [9]	767	0.5

As it results from the review literature [3, 10], electrodes made of carbon in various forms (structures) are one of the most commonly used for the construction of superconductors due to the large specific surface area and low production cost. Additional advantages of this material are good electrical conductivity, chemical stability and a large range of operating temperatures. Despite the properties that pretend carbon materials as ideal for use in supercapacitors, they are also characterized by high resistivity resulting from contact resistance between molecules. In addition, due to the electrochemically inactive part of the pores, the specific capacity of these materials is not high. An effective solution to these problems is the introduction of functional groups or heteroatoms to the surface of the carbonaceous material. They increase ion adsorption and hydrophilicity of the material leading to improved wettability and facilitates fast transport of electrolyte ions inside the micropores. The presence of functional groups on the surface of carbonaceous material can also induce redox reactions, which increases the effective pseudo-capacity of the system.

Recent years have seen rapid development in the field of polymeric materials. A wide range of these materials also included materials for supercapacitors. The polymers used as materials for supercapcitor electrodes must be conductive polymers, and the accumulation of charge takes place in the entire mass of the electrode (and not only on the surface as in the case of EDLC capacitors). Electrical conductivity is obtained in them in oxidation or reduction reactions introducing additional electrons to the polymer chain. However, the problem is the low ionic mobility that reduces the obtained power densities and high resistivity resulting from the smaller distance between the polymer chains than the thickness of the double layer. The solution is, for example, polymer hydrogels [11] or composite electrodes [12, 3].

Eng. Magnetic, Dielectric and Microwave Properties of Ceramics and Alloys   Materials Research Forum LLC
Materials Research Foundations **57** (2019) 75-88   doi: https://doi.org/10.21741/9781644900390-4

The third group of materials for supercapacitors is based on transition metal oxides. From the two above-mentioned groups, they differ in the higher energy density in relation to conventional carbonaceous materials and higher electrochemical stability in relation to polymeric materials. Their special feature is that in the presence of electrolyte they offer intensive redox processes increasing the specific capacitance of the electrodes. Composite electrodes of metal oxides subsidized with various metal oxides, carbon with various structural structures or polymers [3, 13] show particularly desirable properties. An interesting alternative to the typical chemical processes of obtaining composite electrodes based on metal oxides is the use of the magnetron sputtering method that allows obtaining nanostructured electrodes with high specific capacitance [14].

Conclusion

Due to the development of industry and consumer electronics, a significant development in the production of electronic components is observed in parallel. The increasing requirements for capacitor manufacturers to increase the unit capacity and unit energy of capacitors force the development of dielectric materials. Recent years have seen a rapid demand for high power density energy storage devices that would have a high loading and unloading speed. Normal accumulators (or fuel cells) and capacitors do not provide such parameters. This can only be provided by supercapacitors, which undoubtedly represent the future for traditional capacitors. Supercapacitors are increasingly used in parallel with other energy sources, such as fuel cells, for short-term power delivery, which allows for a significant reduction in the overall size of the system. The supercapacitors are also used as continuous power sources in low-power devices, guaranteed power supply systems, and transport.

References

[1] D. A. Nicker, "High Voltage Ceramic Capacitors", Electrocomponent Science and Technology, 1974, Vol. 1, pp. 113-120. https://doi.org/10.1155/APEC.1.113

[2] T. Correia, M. Stewart, A. Ellmore, K. Albertsen, "Lead-Free Ceramics with High Energy Density and Reduced Losses for High Temperature Applications", ADVANCED ENGINEERING MATERIALS 2017, 19, No. 6, 1700019. https://doi.org/10.1002/adem.201700019

[3] G. Wang, L. Zhang, J. Zhang, "A review of electrode materials for electrochemical supercapacitors", Chem. Soc. Rev., 2012, 41, 797–828. https://doi.org/10.1039/C1CS15060J

[4]. M. Ates, "Graphene and its nanocomposites used as an active materials for supercapacitors", J. Solid State Electrochem. 20 (6) (2016) 1509–1526. https://doi.org/10.1007/s10008-016-3189-4

[5] J. Yan, T. Wei, Z. Fan, W. Qian, M. Zhang, X. Shen, F. Wei, "Preparation of graphenenanosheet/carbon nanotube/polyaniline composite as electrode material for supercapacitors", J. Power Sources 195 (9) (2010) 3041–3045. https://doi.org/10.1016/j.jpowsour.2009.11.028

[6] L. Zhang, D. Huang, N. Hu, C. Yang, M. Li, H. Wei, Z. Yang, Y. Su, Y. Zhang, "Threedimensional structures of graphene/polyaniline hybrid films constructed bysteamed water for high-performance supercapacitors", J. Power Sources 342 (2017) 1-8. https://doi.org/10.1016/j.jpowsour.2016.11.068

[7] J. Pedrós, A. Boscá, J. Martínez, S. Ruiz-Gómez, L. Pérez, V. Barranco, F. Calle, "Polyaniline nanofiber sponge filled graphene foam as high gravimetric and volumetric capacitance electrode", J. Power Sources 317 (2016) 35–42. https://doi.org/10.1016/j.jpowsour.2016.03.041

[8] R. Jena, C. Yue, M. Sk, K. Ghosh, "A novel high performance poly (2-methylthioaniline) based composite electrode for supercapacitors application", Carbon 115(2017) 175–187. https://doi.org/10.1016/j.carbon.2016.12.079

[9] Y. Liu, X. Peng, "Recent advances of supercapacitors based on two-dimensionalmaterials", Applied MaterialsToday, 8 (2017) 104–115. https://doi.org/10.1016/j.apmt.2017.05.002

[10] Li Li Zhang, X. S. Zhao, "Carbon-based materials as supercapacitor electrodes", Chem. Soc. Rev., 2009, 38, 2520–2531. https://doi.org/10.1039/b813846j

[11] Soumyadeb Ghosh, Olle Inganäs, "Conducting Polymer Hydrogels as 3D Electrodes: Applications for Supercapacitors", Adv. Mater. 1999, 11, No. 14, pp. 1214-1218. https://doi.org/10.1002/(SICI)1521-4095(199910)11:14<1214::AID-ADMA1214>3.0.CO;2-3

[12] M. Mastragostino, C. Arbizzani, F. Soavi, "Conducting polymers as electrode materials in supercapacitors", Solid State Ionics 148 (2002) 493– 498. https://doi.org/10.1016/S0167-2738(02)00093-0

[13] Jin-HuiZhong, An-Liang Wang, Gao-Ren Li, Jian-Wei Wang, Yan-Nan Ou, Ye-Xiang Tong, "Co_3O_4/$Ni(OH)_2$ composite mesoporousnanosheet networks as a promising electrode for supercapacitor applications", J. Mater. Chem., 2012, 22, 565. https://doi.org/10.1039/c2jm15863a

[14] R. Tummala, R. K. Guduru, P. S. Mohanty, "Nanostructured Co_3O_4 electrodes for supercapacitor applications from plasma spray technique", Journal of Power Sources 209 (2012) 44– 51 https://doi.org/10.1016/j.jpowsour.2012.02.071

Chapter 5

Multiferroics Materials, Future of Spintronics

I.A. Abdel-Latif

Physics Dept., College of Science, Najran University, Najran, P.O. Box 1988, Najran 11001, Saudi Arabia

Advanced Materials and Nano-Research Centre, Najran University, P.O. Box: 1988, Najran 11001, Saudi Arabia

Reactor Physics Dept., NRC, Atomic Energy Authority, AbouZabaal P.O. 13759, Cairo, Egypt

Ihab_abdellatif@yahoo.co.uk

Abstract

Multiferroic materials are a class of new materials where there is a combination between the ferro/antiferroelectricity, the ferro/antiferromagnetism, and the ferro/antiferroelasticity. The most important applications of these materials are their use in spintronics. Progress in developing new materials with new properties suitable for storage media and spin valve transistors is an important step in the field of magnetic materials and their applications. Magnetoresistive random access memory MRAM is one of the applications of the multiferroics materials. In the present chapter, highlights will be focused on the basic concepts of multiferroics science, technology and applications.

Keywords

Spintronics, Multiferroics, Spin Valve Transistor, Magneto-Resistive Random Access Memory, Rare Earth Manganites, Perovskites

Contents

1.   Introduction

Electrons in materials have spin up or down which play an important role in their magnetic properties. From another side, spin could be considered as the "fourth" degree of freedom and it showed a great scientific interest during the last century because of its significance and contribution to physics in general and in particular quantum physics [1], nanomagnetism [2] and spintronics [3]. Materials that are combining ferroelectricity, (ferro) magnetism and (ferro) elasticity are so-called multiferroics where the coexistence of the spontaneous long-range magnetic and dipolar orders are presented in the same time. These attractive materials, which represent very rich properties as well as the fascinating fundamental physics that may allow to use them in potential technological applications particularly in the general area of spintronics. As a result of discovery of giant magnetoresistance [4, 5], spintronics as a branch of electronics appeared and became under the scientist spot, because it offers an option to resolve the existing challenge in magnetic recording and VLSI scaling [6]. Giant magnetoresistance GMR, discovered by A. Fert and P. Grunberg in 1988 and 1989 [3, 4], was the first step of spintronics current to the industrial electronics. Reading of the head of the hard disk drive is one of the important task that excited further desire in the spintronics research that may lead to magnetic tunnel junction [7], spin-transfer torques [8, 9] and non-local spin valves [10]. Particularly, the demonstration of pure spin current injection in non-local spin valves transistor brings in a lot of novel and open questions that have been studied since the very beginning of the new century [11].

The coexistence of magnetic and electric ordering or in other words magnetoelectric coupling is an important physical property and plays an essential role to describe the excitation of the spin waves in an inhomogeneous modulated spin structure where a homogeneous interaction is forbidden by symmetry arguments [12]. The density of the

Eng. Magnetic, Dielectric and Microwave Properties of Ceramics and Alloys   Materials Research Forum LLC
Materials Research Foundations **57** (2019) 89-112   doi: https://doi.org/10.21741/9781644900390-5

free energy as a result of the inhomogeneous magnetodielectric interaction may be determined by

$$\Phi_{me} = -a_x P_x (A_x \partial A_y / \partial y - A_y \partial A_x / \partial y) - a_z P_z (A_z \partial A_y / \partial y - A_y \partial A_z / \partial y),$$

where **P** describes the electric polarization par in this equation, **A** describes the antiferromagnetic vector as a result of the manganese atoms spin in ($GdMnO_3$ & $TbMnO_3$) and $a_{x,z}$ describes the constant of the magneto–dielectric interaction.

The dielectric contribution is given by

$$\Phi_E = -PE + P^2/2\chi_E,$$

where the electric susceptibility is described in this equation by χ_E and the electric field is described by **E**, and minimizing by **P**. The coupling between the spin oscillation and the homogenous a.c. electric field describes the magneto-dielectric interaction which in turn may contribute to the dielectric constant that may cause spontaneous electrical polarization in a modulated magnetic structure.

2.   Ferroelectricity

The microscopic origin of ferroelectricity is much less clear up till now because of many different types and mechanisms that are used to give the interpretation. The basic types of multiferroics [13]

1) Perovskites: d^0 vs d^n

2) "Geometric" multiferroics ($YMnO_3$)

3) Lone pairs (Bi; Pb, ….)

4) FE due to charge ordering

5) FE due to magnetic ordering

When the ferroelectricity is driven by either hybridization or purely structural effects in $BiFeO_3$ as the prototype material one define this material as multiferroics material. Magnetism and FE exist independently, with certain coupling; different sources; different groups of electrons. [14]

When the ferroelectricity is driven by the correlation effects and it is strongly linked to the degrees of freedom of electrons (electronic spin, electronic charge, or orbital-ordering) in rare-earth manganites compounds as prototype materials represent another type of multiferroics. [14]

Eng. Magnetic, Dielectric and Microwave Properties of Ceramics and Alloys   Materials Research Forum LLC
Materials Research Foundations **57** (2019) 89-112   doi: https://doi.org/10.21741/9781644900390-5

- Magnetism and FE exist independently

- FE due to a certain type of magnetic ordering; only in the magnetic state.

2.1   Geometric Ferroelectric

Studying the mechanism of ferroelectricity in certain materials such as $YMnO_3$ [15] is a very interesting topic besides studying the correlation between magnetism and the dielectric property. Anomalous dynamical charges related to the atomic displacement is not displayed for Y nor Mn unlike founded in conventional ferroelectric compounds. The non-centrosymmetric atomic order in the crystal structure occurred as a result of the ionic size effect display the mechanism of the ferroelectricity in this compound that describes the geometric ferroelectricity [12]. Chemical bonds between the cations and anions in such compound play an essential role in the mechanism of the ferroelectricity which called later the chemical bonding effect. The independece of the geometric ferroelectricity on special kinds of cations may lead to expand research in the field of ferroelectric materials. Till date, there is no clear idea describing the nature and the mechanism of the geometric ferroelectricity as well as a clear strategy that enhances their properties. From the reported values of the rare-earth manganites spontaneous polarization [16] (in the range of ~ 5–6 $\mu C/cm^2$) we can say there is no clear trend in the following;

The ferroelectric properties variation in the compounds, the spontaneous polarization, and the transition temperature correlated to the rare-earth elements (ferroelectric Curie temperatures).

2.2   Perovskites

$BaTiO_3$ or $Bb(ZrTi)O_3$ PZT are good example for the well-known ferroelectric materials with the perovskite like structure as shown in Fig.1. On the other hand, perovskite materials have magnetic properties. These materials are composed of transition metal element with the divalent element forming oxides. The partially filled d-shells in a transition metal element is the origin of the magnetic properties, [17] as well as the origin of the ferroelectric properties [18] with an empty d-shell, for instance, Ti^{4+}, Ta^{5+}, W^{6+}. The question is how ferroelectricity occur in these systems? The strong covalent bonds between the transition metal ion based on their empty d-shells and one (or three) oxygens may lead to the off-center shifts of these transition metal ions. When electrons occupy d^n-states in magnetic transition metals may prevent ferroelectricity in magnetic perovskites. And to overcome this problem the mixed d^0 –d^n states are used. As a result of mixing states in the ferroelectrically active ions (d^0-shells and magnetic ions of d^n-

Eng. Magnetic, Dielectric and Microwave Properties of Ceramics and Alloys   Materials Research Forum LLC
Materials Research Foundations **57** (2019) 89-112   doi: https://doi.org/10.21741/9781644900390-5

shell) d^0-ions shift from the centers of O6 octahedra and the polarization and the magnetic order are coexist together.

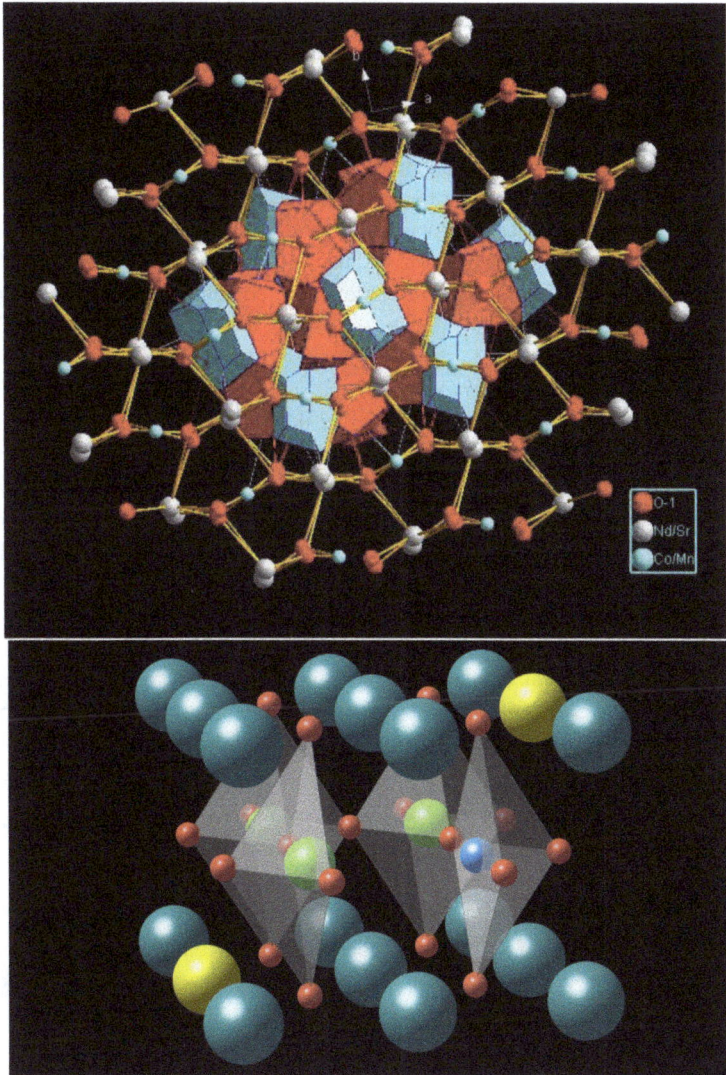

Figure. 1 Different crystal unit cell represenation of perovskites structure [45].

Eng. Magnetic, Dielectric and Microwave Properties of Ceramics and Alloys   Materials Research Forum LLC
Materials Research Foundations **57** (2019) 89-112   doi: https://doi.org/10.21741/9781644900390-5

This point was studied theoretically [19-20], and there were promising results observed by these theoretical calculations but the complete image about this problem still not clear. [21]. The magnetic - ferroelectric coupling is rather weak in this mixed perovskites.

2.3   Ferroelectricity due to lone pairs

As a result of the combination of transition metal oxides such as iron oxides, manganese oxides,...with the trivalent bismuth oxide or the divalent lead oxides, both of the Bi^{3+} and Pb^{2+} ions play the essential role in the formation of the ferroelectricity in these materials where two electrons occupy the $6s$ shell and have no contribution in chemical bonds. These compounds are called lone pairs (dangling bonds) and they express high polarizability because of the ordering of lone pairs in one direction that is making them belong to ferroelectric materials. To improve the ferroelectric properties suggested new combinations such as of $Pb(Zr_xTi_{1-x})O_3$.

2.4   Magnetic multiferroics

In the magnetically ordered state existed in the multiferroics class, ferroelectricity is formed as a result of a particular type of magnetism [22-23]. An example for this class is antiferromagnetic $TbMnO_3$ where magnetic structure changes from 41K to another at 28K discovered and studied. On the other side, electric polarization appears at low temperature where the magnetic field can strongly affect the electric polarization. With other words, we can say that the rotation of the polarization is 90 degrees according to Kimura [22] with the application of a critical magnetic field along a certain direction. The change in field sign may lead to corresponding oscillations in the polarization. This strong magnetoelectric coupling in multiferroics materials is termed as type-II multiferroics. Type-II multiferroics could be divided into two classes in terms of the mechanism of multiferroicity:

First in which the ferroelectricity state is raised as a result of a particular type of *magnetic spiral* and second in which the ferroelectricity state is formed even for the collinear magnetic ordering.

2.5   Ferroelectricity due to charge ordering

Materials in which we found charge-ordering, multiferroicity could be observed. The inequivalent sites and bonds after charge ordering may lead to ferroelectricity in manganites or nickelates [24 -26]. Many compounds with different chemical formula represent this type of multiferroicity such as $RNiO_3$ [24], $TbMn_2O_5$ [25], Ca_3CoMnO_6 [26], $LuFe2O4$ [27].

Eng. Magnetic, Dielectric and Microwave Properties of Ceramics and Alloys   Materials Research Forum LLC
Materials Research Foundations **57** (2019) 89-112   doi: https://doi.org/10.21741/9781644900390-5

3.   Application of multiferroics

In this chapter, we will introduce the most important uses of these multiferroics materials as magnetic storage devices. There are so many devices that could be used as storage media such as multiferroics materials depending upon the use and how often the required data needs to be accessed. [28] For instance, long time storing and infrequently accessed data require hard disk drives with high data density. Studying magnetic and electric transport of perovskite materials is a very important subject where one can determine the magnetoresistance of such materials. Nowadays the capacity of storage is on the range of terabytes (TB) and let us to see how data and information are encoded or stored in the direction of magnetization; (in small areas of a magnetic medium) as we will see later in this work.

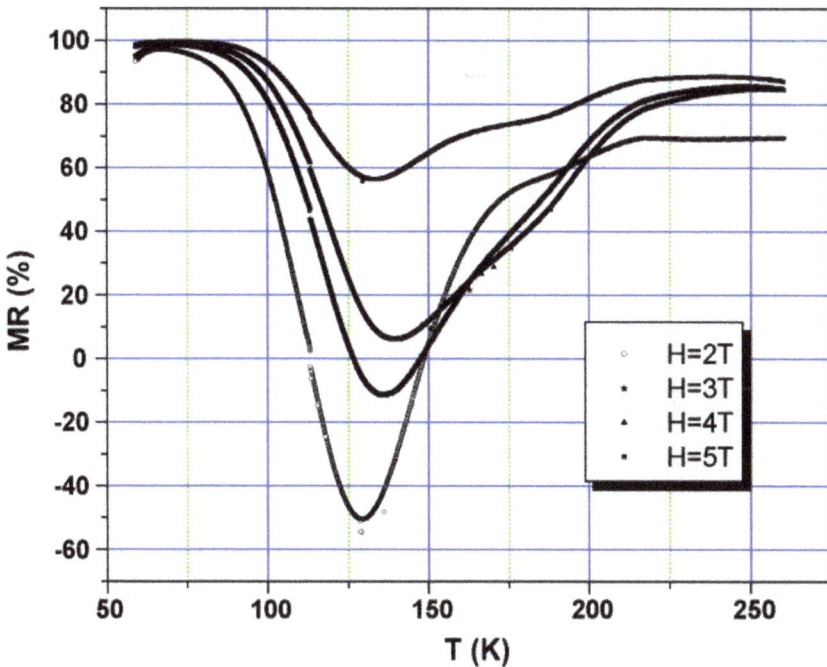

Figure.2 Colossal magnetoresistance of $Sm_{0.6}Sr_{0.4}MnO_3$ at different magnetic fields [30].

Eng. Magnetic, Dielectric and Microwave Properties of Ceramics and Alloys   Materials Research Forum LLC
Materials Research Foundations **57** (2019) 89-112   doi: https://doi.org/10.21741/9781644900390-5

The structural of perovskite and transport properties were investigated intensively in the last few decades [29-39]. These compounds are crystallized with the different crystal structure, where they have the cubic [40-43], orthorhombic [44-53], hexagonal [54-58], rhombohedral [59-64], monoclinic [66-70] crystal system. Based on the structure of any material, physical and chemical properties are determined. ABO_3 perovskite consists of two sides; the A site where A atoms are surrounded by quite distorted 12 oxygen atoms the polyhedral and octahedral side where the oxygen octahedron around the Mn atoms is less distorted. The position of oxygen atom plays a very important role in describing different properties because the interaction between cations in this compound is going through these oxygen atoms. The oxygen atoms around the B-cations in the corner form octahedral sites. There is tilt in these octahedral that cause distortion. This distortion is a very important parameter in the determination of the mutual interaction between the next neighbouring atoms. Abdel-Latif et al., [30] reported in details the role of distortion in the magnetic and electrical transport properties of $Sm_{0.6}Sr_{0.4}MnO_3$.

The first-order approximation of [b] and [c] tilts are defined where we can write them as the following equations;

$$[b] \text{ tilt} \sim (180-\alpha)/2$$

$$[c] \text{ tilt} \sim (180-\beta)/2$$

where α and β are angles between Mn–O_1–Mn and Mn–O_2–Mn, respectively. They calculated tilt angles for $Sm_{0.6}Sr_{0.4}MnO_3$ prepared by themselves and others reported in ref. [68] and compared both values. Moreover, they compared the magnetoresistance in both cases and at different magnetic fields range from 2 T up to 5 T as given in this work. Fig. 2 shows magnetoresistance values at different magnetic fields. Looking at their values we can say that CMR values (97%, 98.7%, 99.3% and 99.7%) are obtained at various magnetic fields (2 T, 3 T, 4 T and 5 T respectively). One of the promising results that are showed by this work is the impressive value of MR at $T = 260K$ (near room temperature). These values are considerable high (69.4% up to 87.3%) in the presence of the magnetic field from 2 T to 5 T, respectively. MR, reported by Dunaevsky in ref. [71] of $Sm_{0.6}Sr_{0.4}MnO_3$ (at magnetic field 2.4 T) but the difference in the condition of synthesis are different from the one obtained by Adel-Latif et al., [30]. Looking at Dunaevsky values of MR at $T = 84.4$ K, in the presence of H= 2.4 T, is 81.7% comparatively small relative to Abdel-Latif MR value between 90.7% and 96.4%. In spite that the structure is the same and the same chemical formula with the same lattice

Eng. Magnetic, Dielectric and Microwave Properties of Ceramics and Alloys   Materials Research Forum LLC
Materials Research Foundations **57** (2019) 89-112   doi: https://doi.org/10.21741/9781644900390-5

constants and the same preparation method (conventional solid-state solution), their MR values have large differences. This difference may be founded as a result of the non-identical thermal treatment during the preparation process that may give rise to the difference in the tilt of octahedral and large difference in MR values. The ortho-ferrimanganites with perovskites like structure where B sites occupied by either iron or manganese are studied by Bashkirov et al., [29]. They showed the role of the different possibilities of cation distributions in the B-site in $SmFe_xMn_{1-x}O_3$ ($x = 0.7, 0.8, 0.9$) using ^{57}Fe Mossbauer spectroscopy at room temperature. Their results proposed several nonequivalent iron/manganese ions distribution in the B-sites based on five sextets experimentally observed when analysis of Mossbauer spectra; each of them represent a different number of the Fe ions in the next nearest neighboring. Their model is so called a nonrandom cation distribution model which is used to find the interpretation of the multi-sextets spectra.

3.1   Magnetoresistive Random Access Memory (MRAM)

Williams Kilburn has practically presented the first electronic random access memory (RAM) in 1947. The electrically charged spots registered on the cathode ray tube was used to produce bits [6-7]. Magnetoresistive random access memory **MRAM** is a technology based on electron spin of multiferroics materials that is used in storing information and now it is part of "Spintronics". MRAM is classified as an ideal memory because they represent a potential combination of the following;

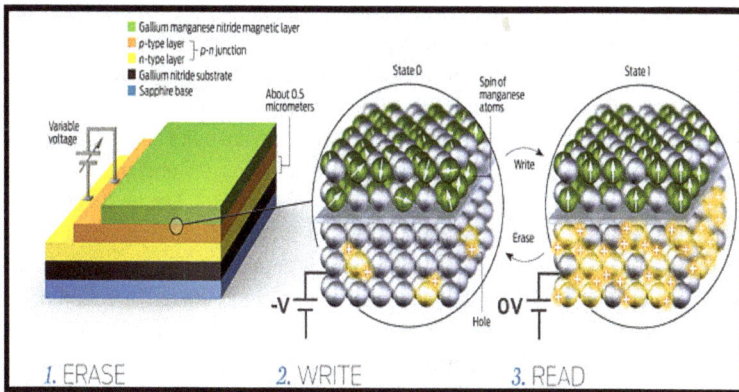

Figure 3 How spin memory works [72].
(http://spectrum.ieee.org/semiconductors/memory/spintronic-memories-to-revolutionize-data-storage (by Salah M. Bedair, John M. Zavada, Nadia El-Masry)).

The density of DRAM with the speed of SRAM and non-volatility of FLASH memory or hard disk, moreover it consumes very low power. Besides all these fantastic characteristics, its resistance to high radiation as well as the possibility to be operated under extreme temperature conditions that makes it is suitable for military and space applications. All the electronic devices with nanoscale dimensions such as memories and processors became denser. Quantum effects had no contributions before but they are now so clear. Searching an effective alternative is the main task of researchers. Spin is an *essential* quantum attribute of electrons that we can use in storing and processing data. The two magnetic states pronounced by spin of electrons; as spin up or spin down could be used to represent the two binary logic in order to store them as a bit.

The progress in the spin-based electronics, or spintronics may open significant capabilities in the field of electronics. It was found that spin is faster and with low energy to be operated which encourages using it. So this new spin transistor will be faster and more powerful than conventional ones. This new technology already exists in the market and is available for use. Magnetoresistive random access memory MRAM is produced as an example of spintronic memory. The MRAM now a day is one of the active subjects for research to develop a new product. We have different kinds of storage used in computers today for example, Dynamic random access memory, or DRAM, Static random access memory, or SRAM. The main characteristic of the DRAM is its high density and low power need but it needs to be constantly refreshed. On the other hand, SRAM could be used in caches. Looking at the flash memory, one can note that it is nonvolatile, at the same time it is quite slow to write, unlike SRAM and DRAM. So one can conclude that the MRAM is attractive and powerful device because of its application in data storage media instead of the other kinds of memory. In principle, bit is represented as charge in a capacitor or as the magnetic state in our device where MRAM stores data according to the electronic spin state (spin up↑ means 0, and spin down ↓ means 1) in the used magnetic materials (for example a ferromagnetic substance) by creating a magnetic alignment in a certain direction, see Fig.3. Let us check the direction of magnetization in an MRAM, we can note that there is fixed magnetization direction belong to the reference layer (or PL) while there is another free layer (FL) varies to store '0' and '1' states [73]. It is clear that the reference layer has a fixed direction because it is fabricated from materials that have a high energy barrier. On the other side, we have a free layer with enough magnetic anisotropy to store the magnetization for certain years.

From the above, we can say the main advantages of the MRAM are the compactness, speed, low-power, and nonvolatile. Based on the application of the MRAM in the computer the management of data processing between the main memory, cache, and disk will be easier and all data directly loaded into its working memory. Moreover, this

Eng. Magnetic, Dielectric and Microwave Properties of Ceramics and Alloys   Materials Research Forum LLC
Materials Research Foundations **57** (2019) 89-112   doi: https://doi.org/10.21741/9781644900390-5

capability may lead to a complete change in the computer architecture that enables working with instant-on. Despite all the above characteristics and advantages of MRAM we still need to solve some problems, for example, the low density of bits and high cost. In the beginning, the change from 1 to 0 required high current and the progress in developing novel materials partially solved this problem but it may be operated with liquid-nitrogen. Of course it is difficult to use them in PC because of the needs for very low temperatures to reduce the write current and this is the main subject of the going research to solve these technical problems.

Let us see how MRAM is working and how to write and read data? The material used in these storage media is a class of semiconductor materials that have magnetic properties such as gallium manganese nitride. Let us see an example of MRAM (spin dependent tunnel junction memory cell) which is shown in Fig. 4. The magnetic row column are used as well as write lines. The change in resistance occurred as a result of the spin-dependent tunnel junction is large and depend upon the electronic spin. The thickness of the tunnel barrier is a very important parameter where it is so small "few atomic layers" and by this way electrons can easily cross through the normally insulating material, giving rise to this change in resistance. In a two dimension array as shown in fig .4, both row and column magnetic write lines enable writing data to be stored in this cell. This writing process required small current to create magnetic fields which in turns cause electronic flipping state (electronic spin up or down) in the spin-dependent tunnel junction storage layer, and hence the change in the junction's resistance. After writing data we need to read this data. To do that we have to know the tunneling current or resistance through the tunnel junction. Further efforts are needed to see next storage media generation with small size and low power consumption taking into consideration the following parameters; Spin-Momentum Transfer, Magneto-Thermal MRAM, and Vertical Transport MRAM. The change in the electron spin as well as the electrical current as a result of the induced magnetic field is a very important parameter that may reduce MRAM write currents. The combination between the magnetic field and ultra-fast heating due to the electrical current pulses played an important role to decrease the required energy to write data. The improved MRAM is the vertical transport MRAM (VMRAM) is another type of MRAM that has high density and in which current is perpendicular to the plane of switching spintronic memory elements. In order to read information, there are three important criteria. The most important parameter is the magnetoresistance (MR) effect which is considered as the key parameter to read data and information in a normal way. Moreover, the difference in voltage between the following states; low and high resistance must be above 0.2 V, to enable reading data and information with high performance.

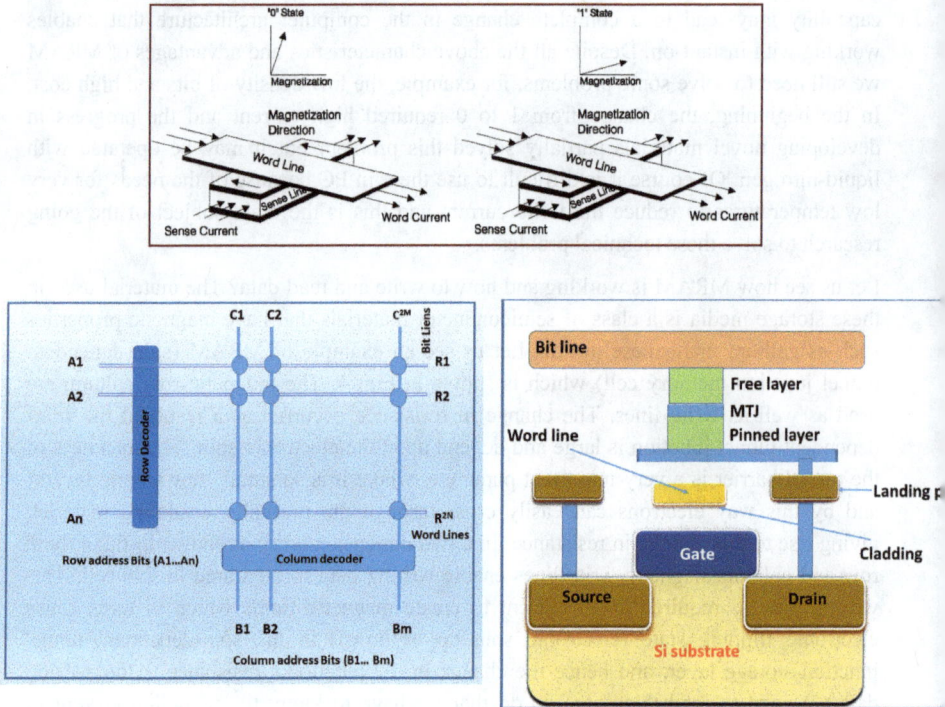

Figure.4 Schematic diagram of classic or conventional MRAM based on spin-dependent tunnel junction memory cells and magnetic row and column write lines [72].

The Mn-based noncollinear antiferromagnetic (AFM) materials is used in spintronics according to Wang et al, [74]. They studied the antiferromagnetic properties of the IrMn-based tunnel junctions and Hall devices. The anisotropic magnetoresistance of IrMn as well as FeMn at room temperature was measured. The investigation found promising spin-orbit effects in AFM as well as spin transfer via AFM spin waves. Their results are an indication to the AFM may be used as an efficient spin current source. From their results, one can say that the AFM metals are a promising candidate for spintronic application.

Theoretical calculations were carried out to design multifunctional materials [75]. Studying the ligand field stabilization and the interaction of a transition metal cation -

Eng. Magnetic, Dielectric and Microwave Properties of Ceramics and Alloys   Materials Research Forum LLC
Materials Research Foundations **57** (2019) 89-112   doi: https://doi.org/10.21741/9781644900390-5

surrounding anions could give us the mechanism of this interaction and present the off-center displacement in the case of the small cation in the common perovskite type of the ferroelectrics for instance; $BaTiO_3$ and $Pb(Zr, Ti)O_3$. These studies were carried out using DFT [76-77]. The tetravalent titanium ion represents the lowest unoccupied energy levels d^0 state which tend to hybridize with O 2p ions [78, 79]. The mutual interaction between an antiferromagnet and a ferromagnet plays a very important role in expressing the so called exchange bias [80].

During the writing process, the in plane magnetization of MRAMs is reaching their limits, while the perpendicular magnetization is promising for use in the near future. According to Song et al., the 8 Mb of STTMRAM in embedded 28 nm logic platform was shown with a high TMR ratio and retention of 10 years [81]. On the other hand, Chung et al. found another promising STT-MRAM, which give a density (4 Gb) at the same time with its compact cell structure (90 nm pitch) [82].

3.2   Spin valve transistor

The progress in electronic devices is one of the scientific research interests. The spin valve transistor is one of the so called spintronic applications (new branch in electronic devices) where the spin-valve transistor is a magneto-electronic device that can be used as a magnetic recording and magnetic sensor. So spin valve transistor is considered as a revolution in the field of magnetic recording and memory devices.

It has a hybrid structure of ferromagnet–semiconductor [80 - 81]. The basic idea in spintronics is to generate spin polarize current from magnetic materials and inject it into other materials as the most successful example given by giant magneto-resistance. As shown in Fig. 1, the basic configuration of GMR is the hybrid multilayer structure of FM/NM/FM (ferromagnetic/non-magnetic /ferromagnetic) [82 - 83], as shown in Fig.5.

Spin transmission depends on the alignment of magnetic moments in the ferromagnets [85]. If a current is passing into a ferromagnet whose majority spin is spin up, for example, then electrons with spin up will pass through relatively unhindered, while electrons with spin down will either 'reflect' or spin flip scatter to spin up upon encountering the ferromagnet to find an empty energy state in the new material. Thus if both the fixed and free layers are polarized in the same direction, the device has relatively low electrical resistance, whereas if the applied magnetic field is reversed and the free layer's polarity also reverses, then the device has a higher resistance due to the extra energy required for spin flip scattering [85].

A magnetic (spin) shape memory effect is one application of the spin-valve-like magnetoresistance properties at room temperature in bulk ferrimagnetic materials. The

Eng. Magnetic, Dielectric and Microwave Properties of Ceramics and Alloys   Materials Research Forum LLC
Materials Research Foundations **57** (2019) 89-112   doi: https://doi.org/10.21741/9781644900390-5

root of this astonishing behavior was in Mn_2NiGa alloy as reported in ref. [86] and it has been studied in details by both of experimental measurements and theoretical calculations as shown in this work. They carried out neutron diffraction and magnetization measurements and as well as ab initio theoretical calculations were done. The direction of the manganese magnetic moments was shown in the soft ferromagnetic cluster inverts when we apply an external magnetic field. The rotation or tilt in the anti-parallel magnetic moment of the manganese atoms at the cluster-lattice interface results in the appearance of the asymmetry in magneto-resistance. [86]

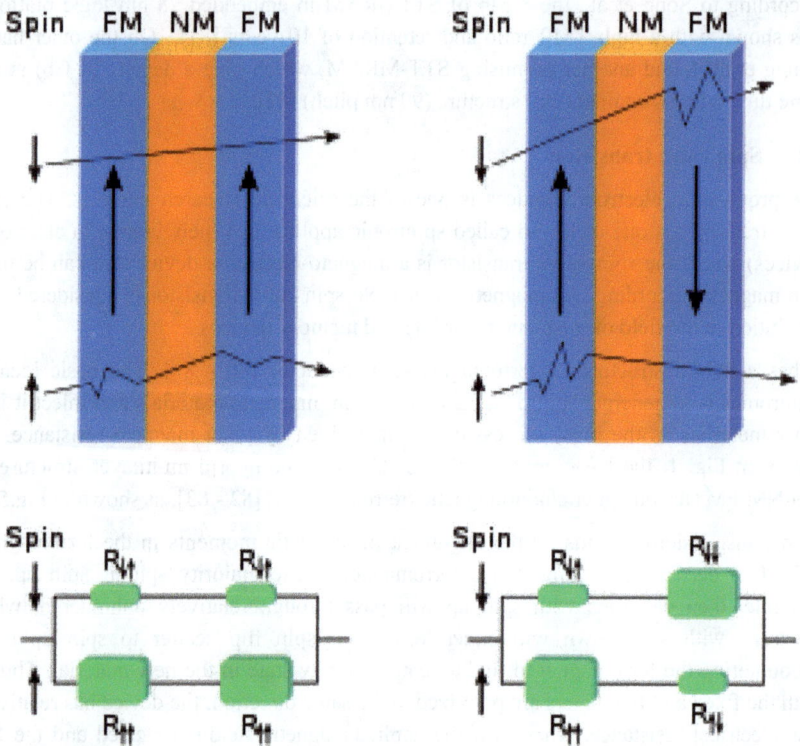

Figure 5 Schematic diagram of multilayers representation of Spin valve transistor. FM: ferromagnetic layer (the direction of magnetization is indicated by arrows), NM: non-magnetic layer. Different spin directions electrons, up and down, scatter differently in the valve [83].

3.2.1 How spin valve transistor works

Spin valves work because of a quantum property of electrons (and other particles) called spin and due to a split in the electronic density of states at the Fermi energy given in the ferromagnetic materials, that gives rise to a net spin polarisation. An electrical current passing through ferromagnetic materials, therefore, carries a charge as well as a spin component. In comparison, a normal metal has an equal number of electrons with different directions of spin (up $\uparrow$ and down $\downarrow$ spins) so, in equilibrium situations, such materials can keep a charge current with a zero net spin component. Anyhow, by passing electric current from a ferromagnetic state into a normal metal it is possible for a spin to be transferred. A normal metal can thus transfer spin between separate ferromagnets, subject to a long enough spin diffusion length [84, 87]. The spin-valve transistor could be defined as a device with three-terminals similar to those in the traditional metallic transistor. The transport of the hot electrons through the spin valve transistor is factionalized somehow to allow these hot electrons to be injected into the interface of the spin valve metal–semiconductor with creating a high Schottky barrier at the emitter of the spin-valve transistor [88]. The energy level diagram of a representative EF is the Fermi level, VBE is the base-emitter bias, VBC is the base-collector bias. [88] as shown in Fig. 7. Spin-valve transistor schematically represented in Fig. 2 (see ref., [87]) with a Si–Pt emitter diode, Si–Au collector diode and a NiFe/Au/Co spin valve; (NiFe = $Ni_{0.81}Fe_{0.19}$). The emitter forward bias injects hot electrons into the spin valve (emitter current IE) as shown in Fig. 6. The organic device based on the organic structure of rubrene ($C_{42}H_{28}$) as an organic semiconductor channel was introduced by B. Li et. al., [89] and they studied the electrical bistability and bias-controlled spin valve effect. They used the ferromagnetic electrodes from the half metallic $La_{0.7}Sr_{0.3}MnO_3$ (LSMO) and Fe. The switching between two states depending on the impedance (low-impedance (ON) state and a high impedance (OFF) state) in Li et al. device are presented. The spin valve effect showed magnetoresistance values up to 3.75% in the ON state. There is no spin valve effect while the device switched to the OFF state. [88].

We propose using the divalent elements doped rare earth manganites of the perovskite-like structure as a ferromagnetic layer with anti-ferromagnetic rare earth manganites layers and comparing its performance with nickel ferrite alloys (Ni-Fe) and Mn_2NiGa alloy in order to get a better performance of the spin valve transistor and to fabricate it. Fabrication of advanced and cheap materials with enhanced properties for spin valve transistor devices to be applied in the field of storage media is an important goal and of scientific interest.

Eng. Magnetic, Dielectric and Microwave Properties of Ceramics and Alloys   Materials Research Forum LLC
Materials Research Foundations **57** (2019) 89-112   doi: https://doi.org/10.21741/9781644900390-5

The magnetically sensitive transistor was originally proposed in 1990 [90] (the spin-valve transistor), and till now under development. Moreover, improvement of the new design of the common transistor appeared in 1940. The spin transistor is one of the results of research on electron spin that naturally express one of two spin states; "spin up" and "spin down".

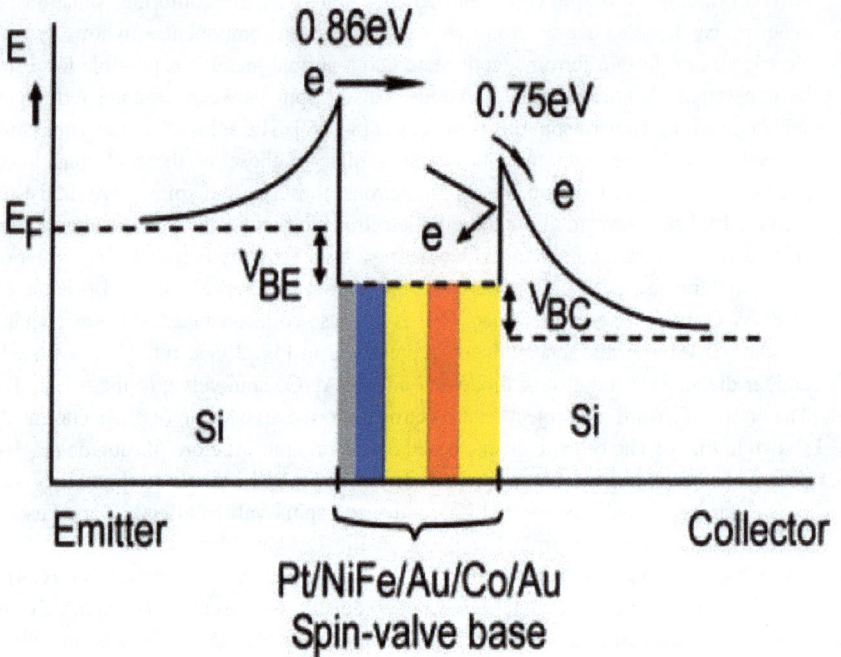

Figure. 6 The schematic energy diagram of the spin-valve transistor showing the Si–Pt emitter and Si–Au collector Schottky barriers and the spin-valve base.

Eng. Magnetic, Dielectric and Microwave Properties of Ceramics and Alloys   Materials Research Forum LLC
Materials Research Foundations **57** (2019) 89-112   doi: https://doi.org/10.21741/9781644900390-5

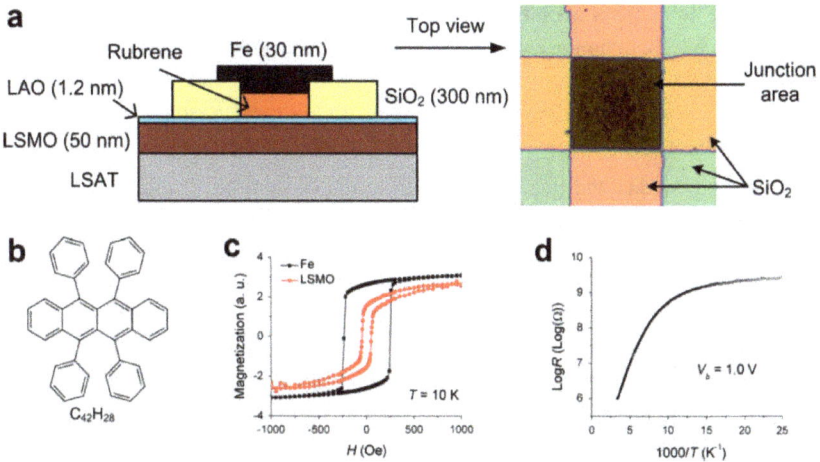

Figure.7 Spin valve representation based on LSMO (a) The schematic device structure. The top view on the right side is showing the optical image of the cross-section. 50 nm rubrene deposited in the center area of 0.2 mm by 0.2 mm and surrounded by SiO2. 30 nm Fe film to be deposited on top of the rubrene layer. (b) Rubrene chemical formula ($C_{42}H_{28}$). (c) Magnetic hysteresis loops representations of LSMO (50 nm) on differetn substrates ; LSAT (0 0 1) substrate and Fe (30 nm) on the glass substrate. (d) Resistance as a function of temperature (OFF state) of the LSMO (50 nm)/LAO/rubrene (50 nm)/Fe (30 nm) junction [88].

Spin transmission depends on the alignment of magnetic moments in the ferromagnets. If a current is passing into a ferromagnet whose majority spin is spin up, for example, then electrons with spin up will pass through relatively unhindered, while electrons with spin down will either 'reflect' or spin flip scatter to spin up upon encountering the ferromagnet to find an empty energy state in the new material. Thus if both the fixed and free layers are polarised in the same direction, the device has relatively low electrical resistance, whereas if the applied magnetic field is inverted and the free layer's polarity also reverses, then the device has a higher resistance due to the extra energy required for spin flip scattering. Spin valves work because of a quantum property of electrons (and other particles) called spin [87, 85]. Due to the split in the electronic density of states at the Fermi energy level in ferromagnetic ordering, there is a net spin polarization. An electrical current passing through a ferromagnet, therefore, shows both charge and a spin component. In comparison, a normal metal has an equal number of electrons of anti-direction of spins (spin up and spin down) so, in equilibrium situations, charge current with a zero net spin component of these materials are expressed. Anyway, when current bypassed from a ferromagnetic metal into a normal metal there is a possibility for transferring spin. d. A normal metal can thus transfer spin between separate ferromagnets, subject to a long enough spin diffusion length.

B. Li [89] has studied the electrical bistability and bias-controlled spin valve effect, in an organic device using rubrene ($C_{42}H_{28}$) as an organic semiconductor channel, where they used the half metallic $La_{0.7}Sr_{0.3}MnO_3$ (LSMO) and Fe as the two ferromagnetic electrodes. The device showed reproducible switching between a low-impedance (ON) state and a high impedance (OFF) state as reported by Li et al. The spin valve effect is shown in the ON state, with magnetoresistance values up to 3.75%. The observed spin valve effect disappears when the device switched to the initial OFF state [89].

The similarity between spin-valve transistor and conventional metallic transistor is that both of could be classified as a three-terminal device [91]. Hot electron transport across the spin valve is reported by Anil Kumar and Lodder [91]. It is employed in order to achieve injection of these hot electrons into the spin valve a metal-semiconductor interface with a high Schottky barrier generated at the emitter of the spin-valve transistor.

Mn_2NiGa alloy has been investigated in ref. [21] using experimental measurements and theoretical calculation showing interesting results of the neutron scattering with magnetization measurements as well as the ab initio theoretical calculations. The orientation of the manganese atom moments in the soft ferromagnetic state inverts according to the applied external magnetic field. The change in the antiferromagnetic order in manganese cluster-lattice interface explained in this work give rise to the observation of asymmetry in magneto-resistance.

The effect of temperature on magnetocurrent was calculated theoretically in the spin valve transistor and explained by Hong and Kumer [88]. They found that the collector current strongly relies on the relative direction of the magnetic moment in the described ferromagnetic metal at different temperatures.

According to Flatte et al., [92] spin-polarized current injection into nonmagnetic semiconductors are theoretically generated of 100% in bipolar transistors with a ferromagnetic base.

Monsma demonstrated [93] the practical integration existed in both common semiconductors and the ferromagnetic semiconductor as spin-valve transistor where a sandwich of the ferromagnetic multilayer existed between two device-quality silicon substrates by means of vacuum bonding. Hot electrons are injected into the spin-valve base somehow to form the emitter Schottky barrier. Si-Pt-Co-Cu-Co-Si devices are achieved at room temperature.

Appelbaum et al. reported the relationship between the spin lifetime and the electron diffusion length in details moreover, they described the scale of coherence in spintronic devices and circuits [94]. As these parameters are many orders of magnitude larger in semiconductors than in metals, semiconductors which make it the most suitable for spintronics. From spin transport measurements in direct-band gap magnetic semiconductors, and neglecting the indirect band gap and non-magnetic semiconductors, the most pronounceable in this work is that silicon, Si, which (in addition to their contribution in electronics) their superiority in spintronics because of their lifetime enhancement and transport length due to low spin-orbit scattering as well as their symmetrical structure (lattice inversion symmetry).

Jansen reported in ref. [95] that spin-polarized tunneling was investigated and their application in the injection and detection of electronic spins in organics and bulk GaAs or silicon. They studied the effect of the electric field control on the electronic spin precession in III-V semiconductors and are dependent on spin-orbit interaction, which makes this proposal insufficient for silicon, the semiconductor attitude. The change in the magnitude of the spin polarization in a quantum well of silicon was demonstrated in this work. The modulation of the spin polarization depends on discrete states in the silicon with a Zeeman spin splitting, in addition to the approach which is also usable for the materials of the weak spin-orbit interaction such as organic and carbon-based compounds.

Van 't Erve et al., found that [96] metals of ferromagnetic ordering are perfect contacts for spin injection and detection, but the problem is the intermediate tunnel barrier required to adjust and tune the large conductivity difference. So that the single-layer graphene successfully showed embrace the traditional meaning of conductivity discrepancy between a metallic state and a semiconductor state in turn of electrical spin injection and detection, where it is of higher order, chemically inert and thermally firm tunnel barrier.

A heteroepitaxial metal-base transistor with perovskite-like structure was given by Yajima et al., [97]. They showed how is the perovskite oxides are candidates for electronic devices due to their large and wide scale of physical and chemical properties in a conventional structure. These new electronic devices are based on the properties of the strongly-correlated electrons. A semiconductor/metal/semiconductor trilayer structure of metal-base transistor is shown, where each layer function described as the common transistor of emitter, base, and collector, respectively. The transistor structure in this work was based on (001)-oriented $SrTiO_3/La_{0.7}Sr_{0.3}MnO_3/Nb:SrTiO_3$ trilayers which were grown by pulsed laser deposition.

Conclusions

From the above, one can conclude that the great progress in materials science and synthesis of new materials with a new crystal structure and with different crystalline size gives us the oppurtunity to discover a wide range of materials with potential applications. Mulitiferroics is one of these class of materials that get much scientific attention because of their applications and uses in the field of electronic devices and there is daily progress that may open the door for improving magnetic storage media and get them in smaller dimensons. From the promising results, metals of ferromagnetic ordering are perfect contacts for spin injection and detection, but on the other side, the problem is in the intermediate tunnel barrier and the large conductivity difference is required to be adjusted and tuned. A heteroepitaxial metal-base transistor with perovskite-like structure showed how is the perovskite oxides are promising candidates for spintronic devices due to their large and wide scale of physical and chemical properties. The spin-polarized tunneling investigation gives us a promising hope because of its application in the injection and detection of electronic spins in different kinds of materials such as organics and bulk GaAs or Silicon.

Acknowledgements

The author is thankful to the Deanship of Scientific Research in Najran University for their financial support NU/ESCI/16/062 in the frame of the local scientific research program support.

References

[1]   J. J. Sakurai, "Modern Quantum Mechanics Revised Edition" (Reading, MA: Anderson-Wesley, (1994).

[2]   S. D. Bader, Colloquium: Opportunities in nanomagnetism, Rev. Mod. Phys., 78(1) (2006). https://doi.org/10.1103/RevModPhys.78.1

[3]   A. Fert, Nobel Lecture: Origin, development, and future of spintronics, Rev. Mod. Phys., 80 (2008)1517. https://doi.org/10.1103/RevModPhys.80.1517

[4]   M. N. Baibich, J. M. Broto, A. Fert, F. N. Van Dau, F. Petroff, P. Eitenne, G. Creuzet, A. Friederich, and J. Chazelas, Giant Magnetoresistance of (001)Fe/(001)Cr Magnetic Superlattices, Phys. Rev. Lett., 61 (1988) 2472. https://doi.org/10.1103/PhysRevLett.61.2472

[5]   G. Binasch, P. Grunberg, F. Saurenbach, and W. Zinn,Enhanced magnetoresistance in layered magnetic structures with antiferromagnetic interlayer exchange, Phys. Rev. B, 39 (1989) 4828. https://doi.org/10.1103/PhysRevB.39.4828

[6]   S. Parkin, X. Jiang, C. Kaiser, A. Panchula, K. Roche, and M. Samant, Magnetically engineered spintronic sensors and memory, Proc. IEEE, 91 (2003) 661. https://doi.org/10.1109/JPROC.2003.811807

[7]   J. S. Moodera, L. R. Kinder, T. M. Wong, and R. Meservey, Phys. Rev. Lett., 74 (1995) 3273. https://doi.org/10.1103/PhysRevLett.74.3273

[8]   M. Tsoi, A. G. M. Jansen, J. Bass, W.–C. Chiang, M. Seck, V. Tsoi, and P. Wyder, Large magnetoresistance at room temperature in ferromagnetic thin film tunnel junctions, Phys. Rev. Lett., 80 (1998) 4281

[9]   J. A. Katine, F. J. Albert, R. A. Buhrman, E. B. Myers, and D. C. Ralph, Current-induced realignment of magnetic domains in nanostructured Cu/Co multilayer pillars, Phys. Rev. Lett., 84 (2000) 3149. https://doi.org/10.1063/1.125752

[10]   F. J. Jedema, A. T. Filip, and B. J. van Wees, Nature, 410 (2001) 345. https://doi.org/10.1038/35066533

[11]   A. Pimenov, A. A. Mukhin, V. Yu. Ivanov, V. D. Travkin, A. M. Balbashov, A. Loidl,Possible evidence for electromagnons in multiferroic manganites, Nature Physics, 2 (2006) 97–100 . https://doi.org/10.1038/nphys212

[12]   T. Toheiet al., Geometric ferroelectricity in rare-earth compounds RGaO3 and RInO3,Phys Rev. B, 79 (2009) 144125. https://doi.org/10.1103/PhysRevB.79.144125

[13]   D. Khomskii, Trend: Classifying multiferroics: Mechanisms and effect, Physics, 2 (2009) 20. https://doi.org/10.1103/Physics.2.20

[14]   Shiqing Deng, Shaobo Cheng, Ming Liu, Jing Zhu., Modulating Magnetic Properties by Tailoring In-Plane Domain Structures in Hexagonal YMnO Films , ACS Applied Materials & Interfaces, (2016). https://doi.org/10.1021/acsami.6b08024

[15]   B. van Aken et al., The Origin of Ferroelectricity in Magnetoelectric YMnO3, Nature Mater., 3(2004) 164. https://doi.org/10.1038/nmat1080

[16]   S Datta, and B. Das, Electronic analog of the electrooptic modulator, Applied Physics Letters, 56 (1990) 665–667. https://doi.org/10.1063/1.102730

[17]   J. B. Goodenough and J. M. Longo, Magnetic and Other Properties of Oxides and Related Compounds, Landolt-Börnstein, Numerical data and Functional Relations in Science and Technology, (4) (1970)

[18]   T. Mitsui et al., Ferroelectrics and Related Substances, Landolt-Börnstein, Numerical data and Functional Relations in Science and Technology, New Series, 16 (1) (1981)

[19] N. A. Hill, Why Are There so Few Magnetic Ferroelectrics, J. Phys. Chem. B, 104 (2000) 6694. https://doi.org/10.1021/jp000114x

[20] D. I. Khomskii, Bull. Am. Phys. Soc. C, 21.002 (2001)

[21] D.I..Khomskii, Multiferroics: Different ways to combine magnetism and ferroelectricity, J. Magn. Magn. Mater., 306 (2006)1. https://doi.org/10.1016/j.jmmm.2006.01.238

[22] T. Kimura et al., Magnetic control of ferroelectric polarization, Nature, 426 (2003) 55. https://doi.org/10.1038/nature02018

[23] N. Hur et al., Electric polarization reversal and memory in a multiferroic material induced by magnetic fields, Nature, 429 (2004) 392. https://doi.org/10.1038/nature02572

[24] D. V. Efremov, J. van den Brink, and D. I. Khomskii, Bond- versus site-centred ordering and possible ferroelectricity in manganites ,Nature Mater., 3 (2004)853. https://doi.org/10.1038/nmat1236

[25] S. W. Cheong and M. V. Mostovoy, Multiferroics: a magnetic twist for ferroelectricity, Nature Mater., 6 (2007)13. https://doi.org/10.1038/nmat1804

[26] Special issue, J. Phys. Condens. Matter, 20 (2008), 434201–434220. https://doi.org/10.1088/0953-8984/20/43/434201

[27] N. Ikeda et al.,Dielectric Relaxation and Hopping of Electrons in $ErFe_2O_4$, J. Phys. Soc. Japan., 69 (2000)1526

[28] Memory with a spin, Spintronic devices that electrically store non-volatile information are promising candidates for high-performance, high-density memories, Nature Nanotechnology, 10 (2015) 185

[29] Sh.Sh. Bashkirov et al., Mössbauer Effect and Electrical Conductivity Studies of $SmFe_xMn_{1-x}O_3$ (x=0.7, 0.8 and 0.9), Journal of Alloys and Compounds, 387 (2005) 70–73. https://doi.org/10.1016/j.jallcom.2004.06.070

[30] I.A. Abdel-Latif et al., The influence of tilt angle on the CMR in $Sm_{0.6}Sr_{0.4}MnO_3$, Journal of Alloys and Compounds, 452 (2008) 245–248. https://doi.org/10.1016/j.jallcom.2007.07.022

[31] Sh.Sh. Bashkirov et al., Crystal Structure, Electric and Magnetic Properties of Ferrimanganite $NdFe_xMn_{1-x}O_3$, IzvestiyaAkademiiNauk. Ser. Fizicheskaya, 67 (2003) 1052.

[32] I.A. Abdel-Latif, S.A. Saleh ,Effect of iron Doping on the Physical Properties of Europium Manganites, Journal of Alloys and Compounds, 530 (2012) 116– 120. https://doi.org/10.1016/j.jallcom.2012.03.079

[33]   K. Bouziane et al., Electronic and Magnetic Properties of SmFe1-xMnxO3Orthoferrites (x = 0.1, 0.2 and 0.3), J. Appl. Phys., 97 (2005) 10504. https://doi.org/10.1063/1.1851406

[34]   I.A.Abdel-Latif et al., Electrical and Magnetic Transport in Strontium doped Europium Ferrimanganites, Journal of Magnetism and Magnetic Materials, 420 (2016) 363–370. https://doi.org/10.1016/j.jmmm.2016.07.016

[35]   I.A. Abdel-Latif et al. Synthesis of novel perovskite crystal structure phase of strontium Doped rare earth Manganites using sol gel method, Journal of Magnetism and Magnetic Materials, 393 (2015) 233–238. https://doi.org/10.1016/j.jmmm.2015.05.078

[36]   M Kh Hamad et al., Effect of cobalt doping in Nd1-xSrxMn1-yCoyO3, Journal of Physics: Conf. Series, 869 (2017) 012032. https://doi.org/10.1088/1742-6596/869/1/012032

[37]   I.A. Abdel-Latif, Study on The Effect of Particle Size of Strontium - Ytterbium Manganites on Some Physical Properties, AIP Conf. Proc., 1370 (2011) 108-115. https://doi.org/10.1063/1.3638090

[38]   I.A. Abdel-Latif, Study on Structure, Electrical and Dielectric Properties of Eu0.65Sr0.35Fe0.3Mn0.7O3, Materials Science and Engineering, 146 (2016) 012003. https://doi.org/10.1088/1757-899X/146/1/012003

[39]   A AYousif, et al., Structure, Electrical and Dielectric Properties of Strontium Europium Ferrimanganites, AIP Conf. Proc., 1370 (2011) 103-107

[40]   Q. Sun, Wan-Jian Yin, J. Am. Chem. Soc., 139 (42) (2017) 14905–14908. https://doi.org/10.1021/jacs.7b09379

[41]   M. Iqbal, J Mater Sci: Mater Electron, 28 (2017)15065–15073. https://doi.org/10.1007/s10854-017-7381-9

[42]   I.A. Abdel-Latif et al., Impact of the Annealing Temperature on Perovskite Strontium Doped Neodymium Manganites Nanocomposites and Their Photocatalytic Performances,Journal of the Taiwan Institute of Chemical Engineers, 75 (2017) 174–182. https://doi.org/10.1016/j.jtice.2017.03.030

[43]   M. B. Salamon and M. Jaime,The physics of manganites: Structure and transport, Reviews of Modern Physics, 73 (2001) 583. https://doi.org/10.1103/RevModPhys.73.583

[44]   V. V. Parfenov, et al., Transport Phenomena of Ferrimanganite Structure Nd0.65Sr0.35FexMn1-xO3, Izv. VyzovPhysica, 10 (2003) 24

[45]   I.A. Abdel-Latif et al., Magnetocaloric Effect, Electric, and Dielectric Properties of Nd0.6Sr0.4MnxCo1-xO3 Composites, Journal of Magnetism and Magnetic Materials, 457 (2018) 126–134. https://doi.org/10.1016/j.jmmm.2018.02.087

[46]   A. Marzouki-Ajmi et al., Journal of Magnetism and Magnetic Materials, 433 (2017) 209–215. https://doi.org/10.1016/j.jmmm.2017.01.097

[47]   I.A. Abdel-Latif et al., Neodymium Cobalt Oxide as a Chemical Sensor, Results in Physics, 8 (2018) 578–583. https://doi.org/10.1016/j.rinp.2017.12.079

[48]   R.I. Zainullina et al., Journal of Alloys and Compounds, 394 (2005) 39–42. https://doi.org/10.1016/j.jallcom.2004.10.032

[49]   K .Das, P. Dasgupta, A. Poddar, I. Das. Significant enhancement of magnetoresistance with the reduction of particle size in nanometer scale. Sci. Rep., 6 (2016) 20351. https://doi.org/10.1038/srep20351

[50]   I. A. Abdel-Latif, et al., Study on Microstructure and Electrical Properties of Europium Manganites, Arab. J. Nucl. Sc. Appl., 44 (2011) 4

[51]   The Nano Particle Size Effect on Some Physical Properties of Neodymium Coblate-Manganites for Hydrogen Storage, I.A. Abdel-Latif, A. Al-Hajary, H. Hashem, M. H. Ghoza and Th. El-Sherbini, AIP Conf. Proc., 1370 (2011) 158-164.

[52]   V. V. Parfenov, I. A. Abdel-Latif , Sh. Sh. Bashkirov On the structure and transport mechanism of Nd0,65Sr0,35Mn1-XFeXO3 solid solution (X=0, 0.2, 0.4, 0.8),, Arab. J. Nucl. Sc. Appl., 40 (2007) 167

[53]   I. A. Abdel-Latif , A. S. Khramov, V. A. Trounov, A. P. Smirnov, Sh. Sh. Bashkirov, V. V. Parfenov, E. A. Tserkovnaya, G. G. Gumarov, Z. Ibragimov, Electrical and Magnetic Properties – Structure Correlation on Nd0.65Sr0.35FexMn1-xO3, Egypt. J. Solids, 29 (2006) 341

[54]   Y. Kumagai, et al., Structural domain walls in polar hexagonal manganites. Nat. Commun, 4 (2013) 1540. https://doi.org/10.1038/ncomms2545

[55]   M. Isobe , et al. Structure of YbMnO3 Acta Crystallogr. C., 47 (1991) 423. https://doi.org/10.1107/S0108270190007995

[56]   H. Ben Khlifa, et al., Journal of Alloys and Compounds, 680 (2016)

[57]   M. Lilienblum, et al., Nature Physics, 11 (2015) 1070–1073. https://doi.org/10.1038/nphys3468

[58]   H Das, A. L. Wysocki1, Y. Geng , W. Wu , C. J. Fennie1, Bulk magnetoelectricity in the hexagonal manganites and ferrites. Nat. Commun., 5 (2014) 2998. https://doi.org/10.1038/ncomms3998

Eng. Magnetic, Dielectric and Microwave Properties of Ceramics and Alloys   Materials Research Forum LLC
Materials Research Foundations **57** (2019) 113-148   doi: https://doi.org/10.21741/9781644900390-6

Chapter 6

The Dielectric Properties of Some Studied Ferrites

Hesham Zaki

Zagazig University, Faculty of Science, Physics Department, Egypt

dakdik2001@yahoo.com

Abstract

In present chapter, AC conductivity and dielectric studies of $Li_{0.5+0.5x}Ge_xFe_{2.5-1.5x}$ (x = 0.0, 0.2, 0.3 & 0.5), $Cu_{1+x}Ge_xFe_{2-2x}O_4$ (x = 0.0, 0.2, 0.3, 0.4), $Cu_{1+x}Ti_xFe_{2-2x}O_4$ (x = 0.0, 0.2, 0.3, 0.4), and $Cu_xFe_{3-x}O_{4+\delta}$ (0.0, 0.4, 0.6, 0.8 and 1.0) spinel ferrites have been covered. The relation of log conductivity and log activation energy for $Li_{0.5+0.5x}Ge_xFe_{2.5-1.5x}$ ferrites showed a straight line between 10^4 Hz to 10^6 Hz. The dielectric properties for Ge substituted and Ti substituted Copper ferrites were explained using the Maxwell–Wagner model.

Keywords

Ge, Ti Substituted Ferrites, AC- Conductivity, Dielectric Properties, Activation Energy

Contents

Eng. Magnetic, Dielectric and Microwave Properties of Ceramics and Alloys   Materials Research Forum LLC
Materials Research Foundations **57** (2019) 113-148   doi: https://doi.org/10.21741/9781644900390-6

1.   Introduction

Spinel ferrite particles in a nano scale have attracted a remarked interest and efforts to investigate their technological applications in many applied applications such as disk recording, microwave industries, electrical devices and so many other applications [1-4].

The introduction of different metallic ions in the spinel structure using a chemical method with a simple reaction can be carried out at relatively low processing temperature may change the magnetic and electric properties considerably. Developing a method of synthesis for optimum properties of ferrites is difficult and complex as well, since most of the properties required for ferrite applications are extrinsic and not intrinsic. However, ferrite is defined by its chemical and crystal structure, as well as requires knowledge and control of different parameter of its microstructures (bulk density, grain and or crystallite size, porosity percentage).

A distinguishable remark of the electrical and magnetic properties in ferrites system depends on the nature of the ions, their charges, and their distribution between tetrahedral (A) and octahedral (B) sites [5].

Many ferrite nanoparticles were proposed as a promising material in many medical applications and biosensors, such as hyperthermia, magnetic drug delivery and magnetic resonance imaging [6-8].

Ferrites are well known for their electrical and magnetic properties with the general structure formula $M^{2+}_{tet} [Fe^{3+}]_{octa} O_4$. By impeding a third metal ion a modification in the distribution of the ions will take place in the spinel structure. The variation in the concentration of third metal ion will change the distribution of Fe^{3+} ion. The electrical, dielectric, magnetic, thermal, and structural properties of the magnetic semiconductor are very sensitive to the chemical composition, type and amount of additives, sintering temperature, and time. The electrical conductivity and the magnetic properties of ferrites are governed by the Fe^{2+}–Fe^{3+} interaction (spin coupling of the $3d$ electrons).

Position of ions on tetrahedral or octahedral site and also the conduction mechanism can be subjected by measuring different properties of spinel ferrites like magnetic and electrical properties.

Some of the applied and basic investigations for the frequency behavior of some ferrites were carried out. These studies of electrical properties of ferrites are interesting and useful for their applications in electronics circuit and apparatus.

Eng. Magnetic, Dielectric and Microwave Properties of Ceramics and Alloys   Materials Research Forum LLC
Materials Research Foundations **57** (2019) 113-148   doi: https://doi.org/10.21741/9781644900390-6

(I) AC conductivity of $Li_{0.5+0.5x}Ge_xFe_{2.5-1.5x}$ (x = 0.0, 0.2, 0.3 and 0.5) ferrites

Frequency and temperature dependences for $Li_{0.5+0.5x}Ge_xFe_{2.5-1.5x}$ (x = 0.0, 0.2, 0.3 and 0.5) ferrites

The (ac) conductivity as a function of frequency for LiGe ferrite is shown in Fig.1 *a, b, c* and *d* respectively. The conductivity is measured between 100 Hz and 1 MHz for different temperatures. It observes from the conductivity relation that it is independent on frequencies at the lower stage and, above a certain frequency, it almost dependent, by increases value with the increase in frequency. Also, the increase in conductivity change rapidly at low temperature and get slower at a higher temperature.

Dispersion in the conductivity of the mentioned ferrites with respect to frequency was explained by Koop's theory [9]. The AC conductivity mechanism can be attributed to that the density of states $N(E_F)$ is finite due to the hopping transport by electrons with energies near the Fermi level and explained by several theoretical treatments [10-12]. Fig. 2 represents the variation of log conductivity with the inverse of temperature for different frequencies, at higher temperatures, the variation in conductivity obey the Arrhenius relation. The activation energy is determined at high temperatures in the straight line region (above *420* K). The evaluated values of conductivity were given in Table. 1 and the insets of Fig. 2. It is observed that the activation energy tends to decrease by the increase in frequency but the conductivity increases a tall temperature by the increase in frequency, and a marked change is prominent at lower temperatures. The results can be explained using the hopping model assumed by Pike [2], which explained the AC conductivity results for many oxides and materials in the amorphous state.

Table 1: The variation of activation energy versus frequency for each composition of $Li_{0.5+0.5x}Ge_xFe_{2.5-1.5x}$ ferrites.

f (Hz)	$\tilde{W}$ (eV)			
	x = 0.0	x = 0.2	x = 0.3	x = 0.5
10^2	0.51	0.42	0.22	0.25
10^3	0.51	0.42	0.22	0.25
10^4	0.51	0.38	0.22	0.25
10^5	0.43	0.31	0.20	0.23
3×10^5	0.37	0.25	0.19	0.22

Eng. Magnetic, Dielectric and Microwave Properties of Ceramics and Alloys   Materials Research Forum LLC
Materials Research Foundations **57** (2019) 113-148   doi: https://doi.org/10.21741/9781644900390-6

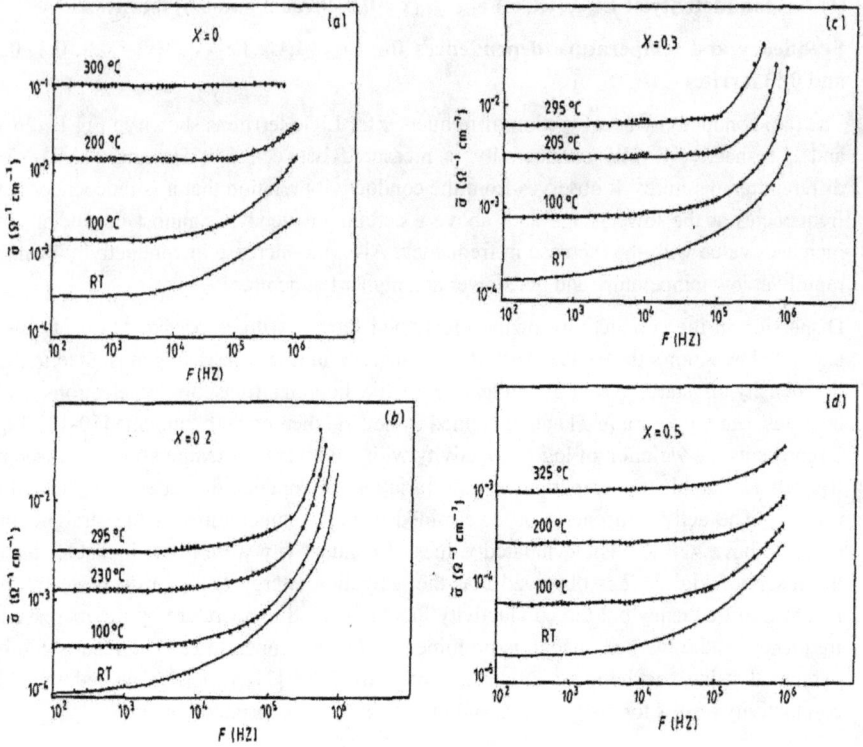

Fig.1 Plots of AC conductivity against frequency at different temperatures for
$Li_{0.5+0.5x}Ge_xFe_{2.5-1.5x}$ (x = 0.0, 0.2, 0.3 and 0.5) ferrites.

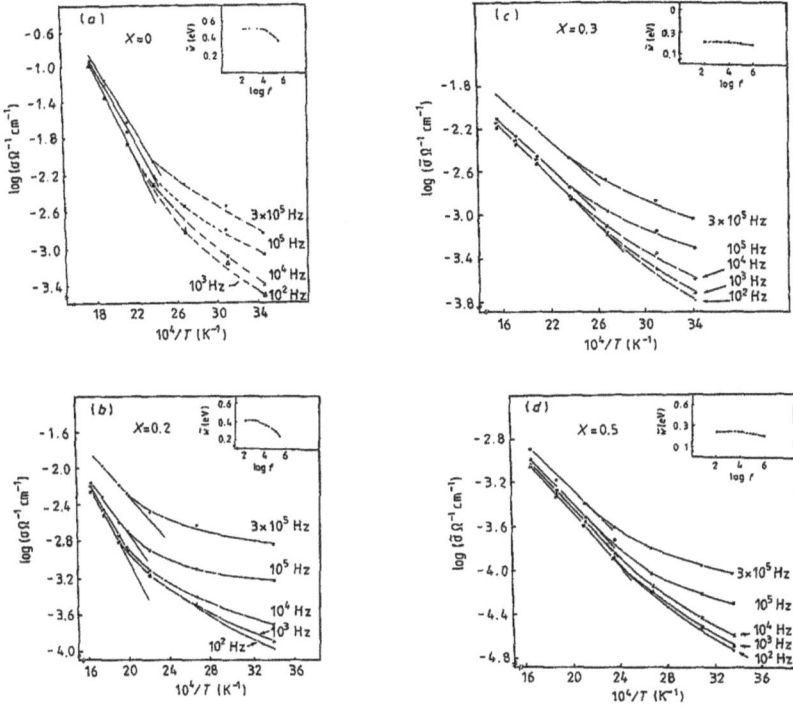

Fig.2 Variation of log $\tilde{\sigma}$ as a function of T^{-1} at different frequencies for $Li_{0.5+0.5x}Ge_xFe_{2.5-1.5x}$ (x= 0.0, 0.2, 0.3 and 0.5).

Fig. 3 shows the variation of log conductivity and log activation energy for the proposed composition of $Li_{0.5+0.5x}Ge_xFe_{2.5-1.5x}$ ferrite. The relation showed a straight line within the frequency range between 10^4 Hz to 10^6 Hz at R.T. The calculated values of S (exponent factor) and v_{ph} (at f = 10^5 Hz) for all composition are shown in Table 2. The relation between composition and v_{ph} is given in Fig. 4.

Eng. Magnetic, Dielectric and Microwave Properties of Ceramics and Alloys   Materials Research Forum LLC
Materials Research Foundations **57** (2019) 113-148   doi: https://doi.org/10.21741/9781644900390-6

Table 2: The calculated values for S and v_{ph} (at f = 10^5 Hz) for each composition

X	S	v_{ph} (Hz)
0.0	0.45	9.05×10^8
0.2	0.80	3.05×10^{14}
0.3	0.70	3.88×10^{11}
0.5	0.60	1.38×10^{10}

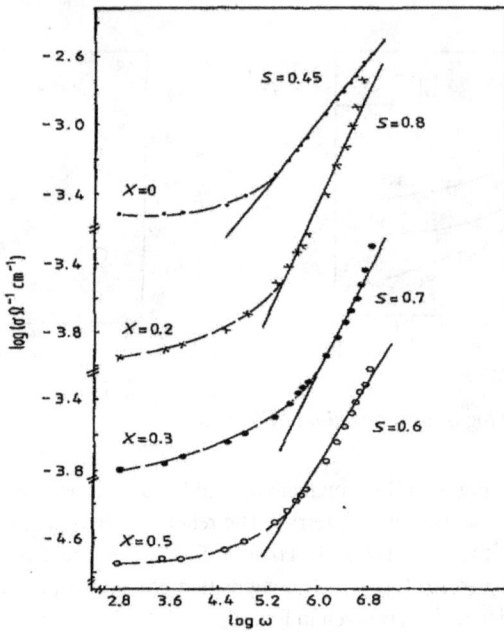

Fig.3 The relation between log $\bar{\sigma}$ and log ω.

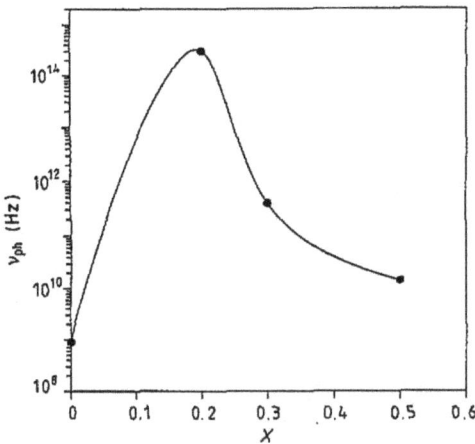

Fig. 4 The relation between v_{ph} and the composition.

2.   The loss tangent

Fig. 5 shows the variation loss tangent ($tan\delta$) against frequency for samples with x = 0.0, 0.2, 0.3 & 0.5. The measurements were carried out at room temperature with a frequency range from 100 *Hz* up to 1 MHz. The estimated values for the dielectric constant and the dielectric loss against conductivity at room temperature are shown in Fig. *6.* A dispersion behavior appeared for all compositions in $tan\delta$ against frequency curve. For the samples with x= 0.2, 0.3 and 0.5 different peaks position named ($tan\delta$)$_{max}$ at frequencies 5.5 x 10^4, 5 x 10^4 and 1.5 x 10^5Hz, respectively. But $tan\delta$ gives a maximum in the frequency range (2 up to 6) x 10^4Hz, for sample x = 0.0. Also, a minimum value for the dielectric constant ε' is found within the corresponding frequency value. The observation of $tan\delta$ peaks can be explained owing to the dispersion of dielectric constant. These maxima may be attributed to the fact that the dispersion in ε' is large enough to reflect a noticeable peak in $tan\ \delta = \varepsilon''/\varepsilon'$, where ε'' is the imaginary part of the complex dielectric constant that describes the dissipation energy. This maximum in $tan\delta$ reflects the rapid decrease in ε' as the increase in frequency due to the dispersion and *a* slowly decreases in ε'', as is observed in Fig. 6. Also a likely explanation for the occurrence of peaks can be given according to the relation between the dielectric behavior and the conduction mechanism of the ferrites. The conduction mechanism mainly depends on the hopping motion of the electron between Fe^{2+} and Fe^{3+} over the octahedral site in the ferrites. Then, $tan\delta_{max}$ can

Eng. Magnetic, Dielectric and Microwave Properties of Ceramics and Alloys   Materials Research Forum LLC
Materials Research Foundations **57** (2019) 113-148   doi: https://doi.org/10.21741/9781644900390-6

be attributed to the resonance effect, which can be observed when the jump frequency between the ferric and ferrous ions is nearly equal to that of the external electric field.

(II) Effect of tetra ionic substitution on the dielectric propertiesof Cu-ferrite:

Frequency dependence for $Cu_{1+x}Ge_xFe_{2-2x}O_4$ and $Cu_{1+x}Ti_xFe_{2-2x}O_4$ systems

The ac conductivity against Frequency at room temperature for the two systems ($Cu_{1+x}Ge_xFe_{2-2x}O_4$ and $Cu_{1+x}Ti_xFe_{2-2x}O_4$ with $0 \leq x \leq 0.4$) was measured. The variation of the AC conductivity with frequency is represented in Fig. 7a (between 10^2 and 10^6 Hz) at room temperature for Cu- ferrite (as a base ferrite). Another representation is shown in Fig. 7 b and c for the doped ferrite systems (system 1 for Cu-Ge and system 2 for Cu-Ti). The AC conductivity reflects an increase with frequency for all samples. Below 10^4 Hz, the conductivity seems to be frequency independent and hardly increases with frequency. But, after 10^4 Hz, it obviously increases with frequency for the two systems. In the high-frequency range, A saturation was reached for the composition with x = 0.4 ($Cu_{1.4}Ti_{0.4}Fe_{1.2}O_4$). As we know the dielectric constant ε (as a complex) is represented as:

$$\varepsilon = \varepsilon' - i\varepsilon'' \tag{1}$$

Where ε' and ε'' describe the stored energy and dissipated energy respectively. Fig. 7 (a-c) depict the frequency dependence for both the dielectric constant (ε') and the dielectric loss (ε'') for the two systems, at room temperature. It is clear that ε' and ε'' for both systems show a decrease with increasing frequency, however, ε'' decreases faster than ε' over the same rage of frequency. It is clear that in the high-frequency range (above 10^5 Hz) both ε' and ε'' become closer in both systems.

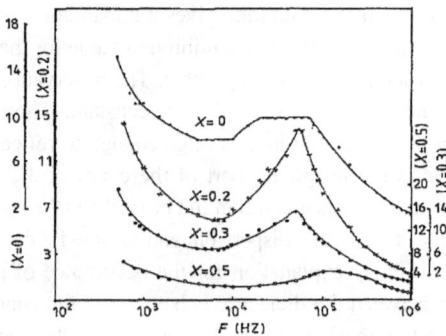

Fig. 5 The dependence on the loss tangent on the frequency.

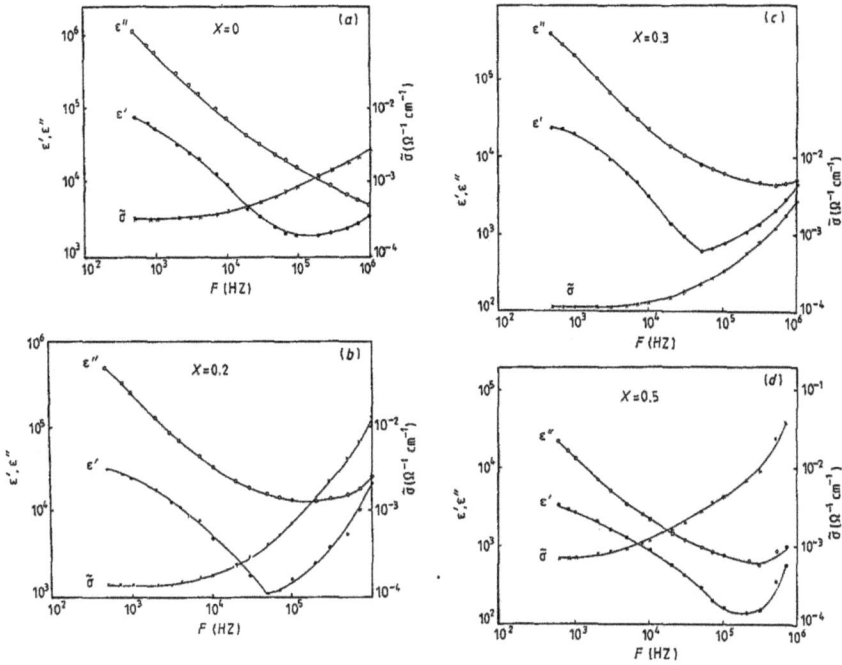

Fig. 6 Plots of σ, ε' and ε'' against frequency at room temperature for $Li_{0.5+0.5x}Ge_xFe_{2.5-1.5x}$ (x= 0.0, 0.2, 0.3 and 0.5).

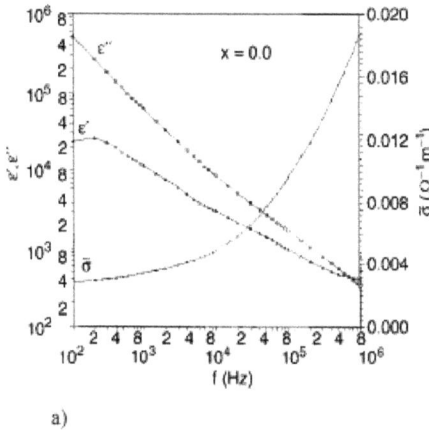

a)

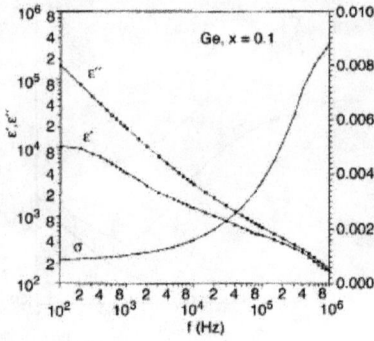

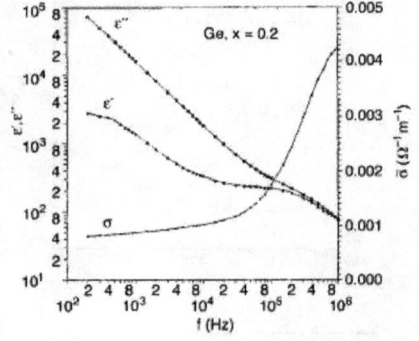

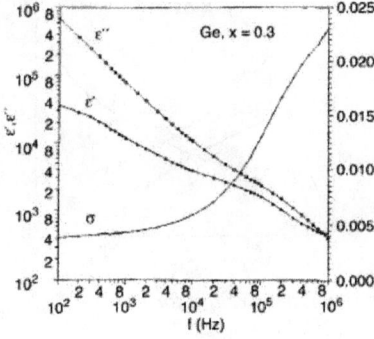

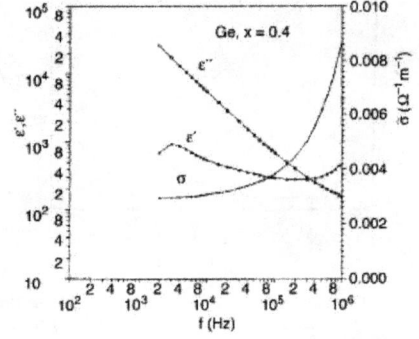

Eng. Magnetic, Dielectric and Microwave Properties of Ceramics and Alloys   Materials Research Forum LLC
Materials Research Foundations **57** (2019) 113-148   doi: https://doi.org/10.21741/9781644900390-6

b)

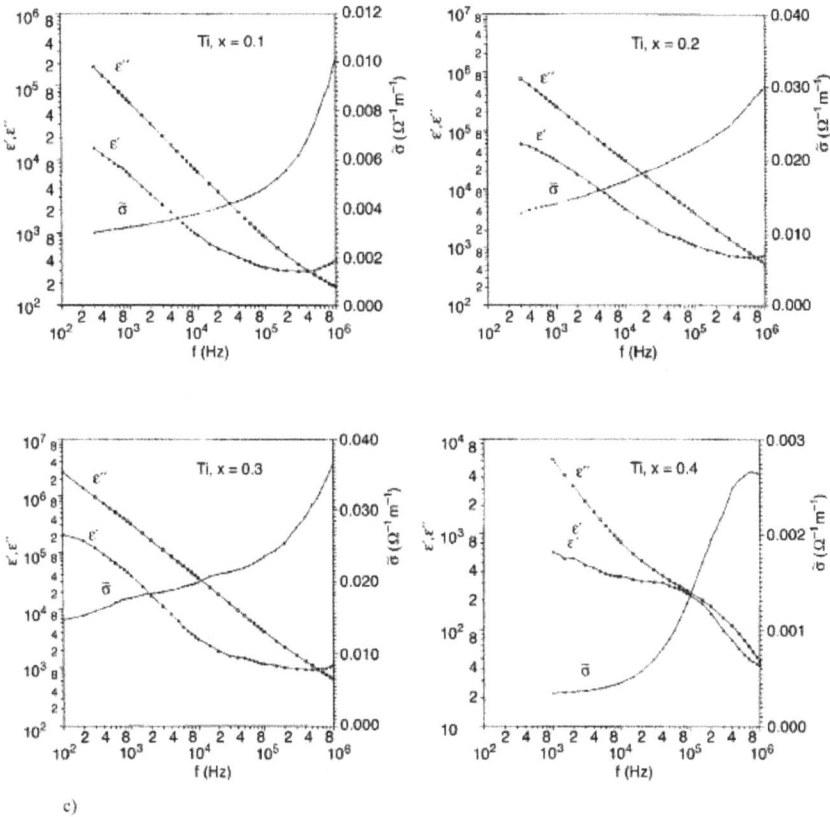

Fig. 7 The variation of the ac conductivity σ, the dielectric constant ε', and the dielectric loss ε'' against frequency for a) $CuFe_2O_4$, b) $Cu_{1+x}Ge_xFe_{2-2x}O_4$, c) $Cu_{1+x}Ti_xFe_{2-2x}O_4$.

In the studied systems, Fig. 8 (a-c) represent the loss tangent against frequency. The loss tangent shows a decreasing value exponentially as frequency increase in the range of 10^2 to 10^6 Hz. Tan δ becomes frequency independent and approaches unity In the high-frequency region which may reflect the fact that within this frequency region $\varepsilon' \approx \varepsilon''$. The samples show strong temperature dependence at low frequencies, as shown in Fig. 8 (a-c) for both ferrite systems ($CuFe_2O_4$, Cu–Ge, and Cu–Ti), respectively. Dielectric loss (*tan δ*) starts to appear at a certain frequency (f_{st}) for each composition, [13]. Before f_{st}, the

Eng. Magnetic, Dielectric and Microwave Properties of Ceramics and Alloys   Materials Research Forum LLC
Materials Research Foundations **57** (2019) 113-148   doi: https://doi.org/10.21741/9781644900390-6

compositions have pure resistance behavior with a constant conductivity, which means ε' approached zero and *tan δ* approached infinity. After f_{st}, the samples have a dielectric behavior and *tan δ* can be determined for both systems. As temperature increases, f_{st} increases gradually which clear in Fig. 8. The given data for the dielectric properties was analyzed according to the Maxwell–Wagner model [14]. The model stated that the ferrite is supposed to be inhomogeneous with two regions, one is thin and poorly conducting grains which separated by a thick and conducting layer. As the frequency is low, the thin layers which are poorly conducting give rise to a very high permittivity, but at higher frequencies they behave as a capacitive short-circuit. Otherwise, a dispersion of dielectric constant and ac conductivity is found in the intermediate frequency range. The electrical properties of the bulk material seem to be suppressed by the layers at low frequency to a certain extent. At high frequency, the measuring results are directly related to values for the conductivity and permittivity of the bulk material. AC conductivity and permittivity are given by the dispersion formula [15]:

$$\tilde{\sigma}(\omega) = \sigma_h + \frac{\sigma_l - \sigma_h}{1 + (\omega\tau)^2} \tag{2}$$

$$\varepsilon'(\omega) = \epsilon_h' + \frac{\epsilon_l' - \epsilon_h'}{1 + (\omega\tau)^2} \tag{3}$$

Where σ_l, σ_h, ϵ_l' and ϵ_h' denotes the low- (L) and high (h)-frequency saturation values of the conductivity and the permittivity of the material of the two layers. $\omega = 2\pi f$, where f is the frequency and τ is the relaxation time. It is the mean jump time from lower to upper state [16]. It should be emphasized that σ_l, σ_h, ϵ_l' and ϵ_h' are related to the intrinsic properties of the material.

The AC conductivity was evaluated as a function of frequency, for both systems. The best agreement was produced between both calculated and the experimental values of conductivity by taken the relaxation time (τ) as a variable, which given by the relation [17]:

$$\tau = \tau_o \exp\left(\frac{W}{kT}\right) \tag{4}$$

Where τ_0 is the atomic vibration period order of an atomic vibrational period.

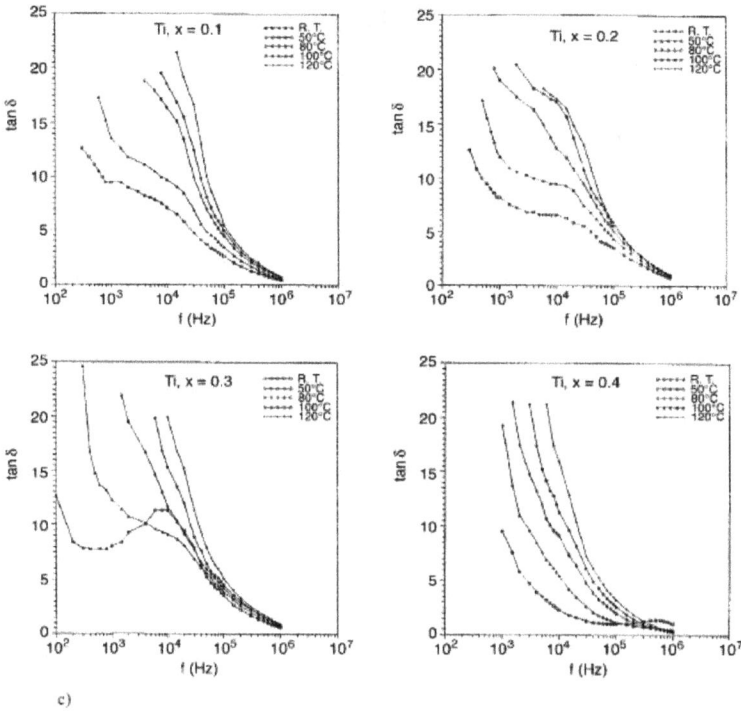

Fig. 8 Frequency dependence of the loss tangent tan δ of a) Cu ferrite, b) Cu–Ge ferrite, c) Cu–Ti ferrite.

For the system of Cu-Ti (x=0.4) a good agreement is found with the value of relaxation time (τ) equal 1.85×10^{-6} s at R.T. The fit is applied for Cu-Ge system (x = 0.2) because the high-frequency saturation is not observed for this system, with x = 0.4, where a saturation is almost reached. Accordingly, the best fit is found to be about 5×10^{-7} s at R.T. The relation between relaxation time and temperature is strong enough to take the relaxation times $\tau_1 = 5 \times 10^{-7}$ s and $\tau_2 = 1.85 \times 10^{-6}$ s for both systems. The calculated hopping rates (g) for both systems, are given by $2 \times 10^{6}\,\text{s}^{-1}$ and $5.4 \times 10^{5}\,\text{s}^{-1}$, respectively. The estimated values of the conductivity are shown as the dashed lines in Fig. 9 a and b for x = 0.2 for Cu-Ge (system 1) and x = 0.4 for Cu-Ti (system 2).

The hopping model state that the AC conductivity is usually frequency independent as the applied field frequency is less than the loss peak frequency, denoted by f_L, as shown in Fig. 9, where loss peak frequency gives the lowest effective jump frequency of the

Eng. Magnetic, Dielectric and Microwave Properties of Ceramics and Alloys   Materials Research Forum LLC
Materials Research Foundations **57** (2019) 113-148   doi: https://doi.org/10.21741/9781644900390-6

ferrite. But the saturation at high frequency presents when the applied field frequency equals f_h. In the dispersion region, between f_L and f_h (as present in Fig. 9) the power law of the frequency is applied because of the strong increase in conductivity [18].

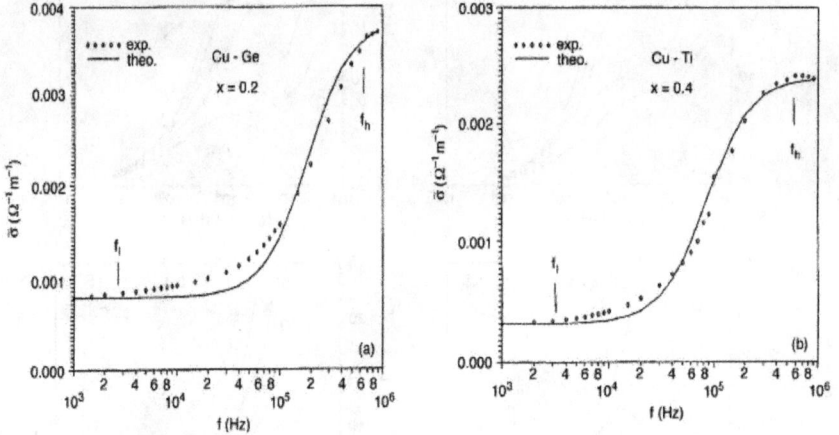

Fig. 9 Comparison between the experimental values and the calculated values of σ as a function of frequency for: a) $Cu_{1.2}Ge_{0.2}Fe_{1.6}O_4$ at τ = 5 × 10^{-7} s, b) $Cu_{1.4}Ti_{0.4}Fe_{1.2}O_4$ at τ = 1.85 × 10^{-6} s.

3. Temperature dependence for Cu-Ge and Cu-Ti ferrite system

The AC conductivity was measured within the range of temperature between 300 K and 773 K, at four different frequencies 10^2, 10^3, 10^4, and 10^5 Hz for both systems and shown in Fig. 10 a, b, and c. This figure shows a semiconducting trend as seen for both ferrites. Regarding this figure, the conductivity is almost frequency independent at high temperatures, while at low temperatures dispersion was observed, the AC conductivity was increased as frequency was increasing in this dispersion region. The conductivity is seen to be a weak temperature dependent in the higher range, of frequency while it shows strong temperature dependence in the low frequency range. The values of activation energy for each composition for each system, below the temperature of 480 K, were calculated at four different frequencies according to the relation [16]:

$$\tilde{\sigma} = \sigma_0 \exp\left(\frac{-E_\sigma}{kT}\right) \tag{5}$$

Eng. Magnetic, Dielectric and Microwave Properties of Ceramics and Alloys   Materials Research Forum LLC
Materials Research Foundations **57** (2019) 113-148   doi: https://doi.org/10.21741/9781644900390-6

where σ_0 is a temperature-dependent term and E_σ represents the activation energy of the conduction mechanism and the other parameters have their usual meaning.

Table 3 shows the calculated values of the activation energy. In general, it can be observed that there is a decrease in activation energy with the increase in frequency for the studied ferrites system to some extent. Cu–Ti ferrite system shows that activation energy ranged between 0.21 and 0.3 eV is lower than that of Cu–Ge ferrite system (0.27–0.39 eV) as shown in table 3 and this may be due to the difference in screening produced by each tetravalent ion. The difference in ionic radius of Ge^{4+} ion (0.44 nm) and the ionic radius of Ti^{4+} ion (0.68 nm) responsible for the screening potential (which is inversely proportional to the ionic radius [19]). It is observed that the screening behavior for Ge^{4+} ion is greater than that of Ti^{4+} ion. Accordingly, the Cu–Ti ferrite activation energy tends to be lower than the corresponding activation energy for Cu–Ge ferrite.

Table 3: The variation of ac activation energy vs. frequency (for T < 480 K).

frequency (Hz)	E (eV)								
	$CuFe_2O_4$ $Cu_{1+x}Ge_xFe_{2-2x}O_4$					$Cu_{1+x}Ti_xFe_{2-2x}O_4$			
	$x = 0.0$	0.1	0.2	0.3	0.4	0.1	0.2	0.3	0.4
10^2	0.39	0.39	0.32	0.33	0.31	0.275	0.265	0.255	0.30
10^3	0.39	0.39	0.32	0.33	0.31	0.270	0.250	0.24	0.30
10^4	0.36	0.36	0.29	0.31	0.31	0.255	0.240	0.230	0.290
10^5	0.30	0.33	0.27	0.29	0.29	0.240	0.225	0.210	0.250

The observed low values for activation energy may be explained due to the paths should be taken for the DC conductivity must be the easiest path between the ions and these paths must include some jumps for which R, the distance between the ions, is large which is not important in the AC conduction which means low activation energy for the AC conduction.

Eng. Magnetic, Dielectric and Microwave Properties of Ceramics and Alloys   Materials Research Forum LLC
Materials Research Foundations **57** (2019) 113-148   doi: https://doi.org/10.21741/9781644900390-6

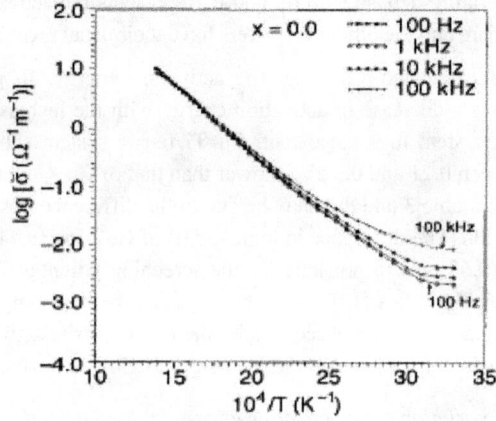

a)

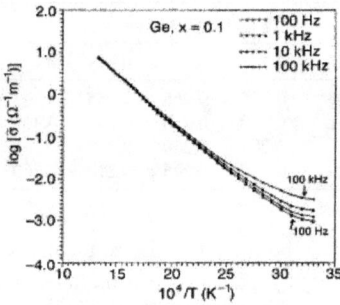
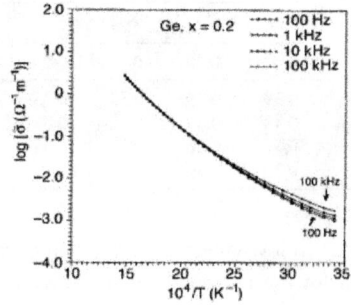

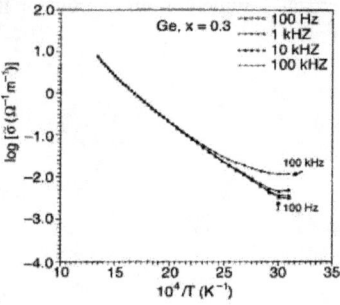
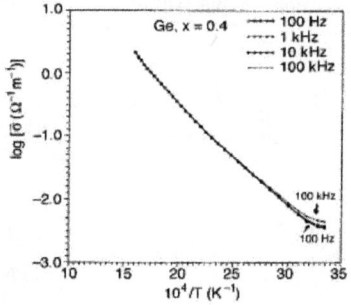

b)

Eng. Magnetic, Dielectric and Microwave Properties of Ceramics and Alloys   Materials Research Forum LLC
Materials Research Foundations **57** (2019) 113-148   doi: https://doi.org/10.21741/9781644900390-6

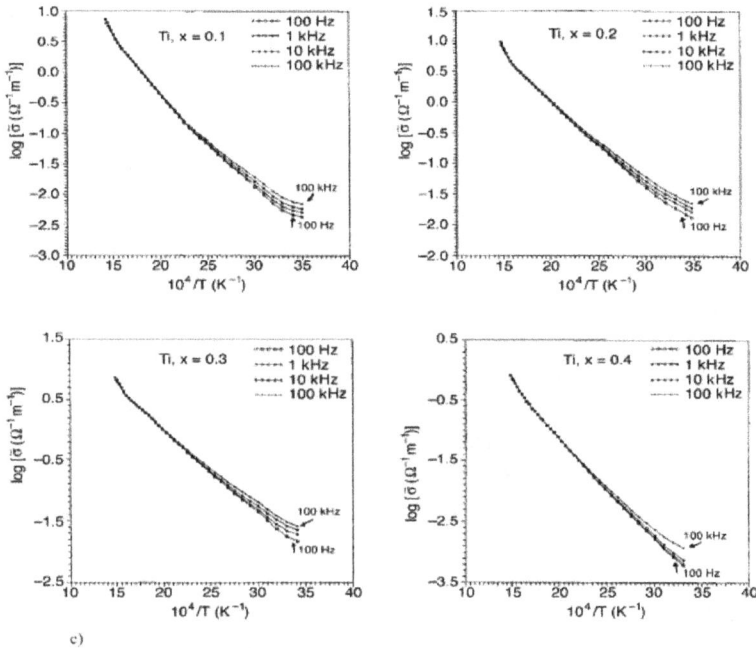

Fig. 10 The temperature dependence of AC conductivity at four frequencies for a) Cu ferrite, b) Cu–Ge ferrite, c) Cu–Ti ferrite.

The dependence of the dielectric constant ε' at different frequencies for based Cu ferrite, Cu–Ge and Cu–Ti systems against temperature is shown in Fig. 11 a, b, and c, respectively. It is observed that the increase of dielectric constant against temperature until a maximum is reached and then no further increase but a sharp decrease. As the applied frequency is increased the temperature at which a drop in dielectric constant (ε') occurs will shift towards lower temperatures. This behavior was seen in Li–Ti ferrite system [20]. Also, an abnormal behavior was observed in Cu–Mn, Cu–Ni and Cu–Zn ferrites [21], where ε' increases with increasing temperature having a maximum peak that shifts towards higher temperature with increasing frequency. According to Rezlescu [21] a suggested model depends on the contributions of two types of carriers (n) and (p) to polarization depend with temperature was based to explain the abnormal behavior of the dielectric constant against temperature. But the model is not valid for the given results for many reasons. The first one is that the maximum peak, in dielectric constant in the result

Eng. Magnetic, Dielectric and Microwave Properties of Ceramics and Alloys   Materials Research Forum LLC
Materials Research Foundations **57** (2019) 113-148   doi: https://doi.org/10.21741/9781644900390-6

we have, shifts toward low temperature by the increase in frequency (opposite to their results). Secondly, the model state that the p-carriers must increase with the temperature where the density our results for the p-carriers [22] is found to be decreased or even constant at the temperature related to the maximum peak of ε'. So the result which we got may be explained as follows:

As Iwavachi [23] has pointed out that the correlation between the conduction mechanisms hopping process and the behavior of dielectric in ferrites is very strong. Qualitatively the behavior can be explained by the supposition theory where the mechanism of the polarization process in ferrites is like that of the conduction process. Due to the exchange process ($Fe^{2+} \Leftrightarrow Fe^{3+} + e^{-}$), one obtains local displacements of electrons in the direction of the applied electric field. The polarization in ferrite was determined by these displacements. As we know that the polarization effect is reducing the field inside the medium. So, the decrease in polarization (the dielectric constant ε') by the increase in frequency is related to the fact that, beyond a certain frequency of the electric field, the electronic exchange between ferric and ferrous ions cannot follow the alternating field [20, 24 and 25].

The increase in temperature has a tendency to reduce the charge of p-carriers density, as given by the measurements of thermoelectric power [22], despite the fact that the temperature does not directly influence the electronic polarization. Hence the electronic contribution will increase the polarization. For that, the dielectric constant will increases by increasing temperature firstly. In addition, the relaxation time is reduced by the rise in temperature [26]. The dielectric constant will increase by the increase in temperature $[\epsilon'(\omega) = \epsilon'_h + \frac{\epsilon'_l - \epsilon'_h}{1+(\omega\tau)^2}, \tau = \tau_o \exp(\frac{W}{kT})]$.

Moreover, at a certain temperature, the dielectric constant may be drop. This drop may be related to the effect of disorientation of the random thermal motion as the temperature increased. But, according to the relation between the dielectric constant and the frequency (inverse) the higher the frequency the less the electronic exchange contribution to the polarization. Hence the drop in dielectric constant could shift towards lower temperature by the increase in frequency, which is opposite to the prediction of the Rezlescu model.

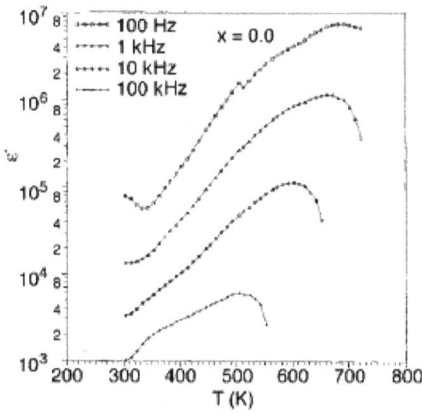

a)

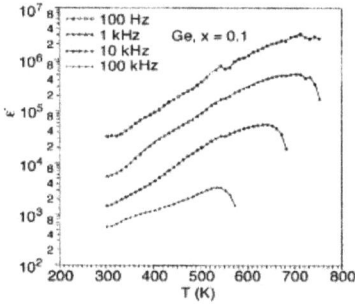
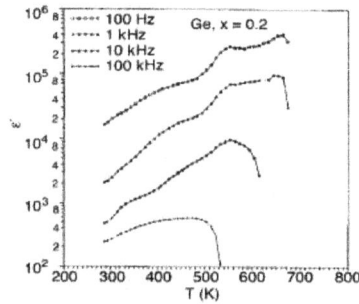

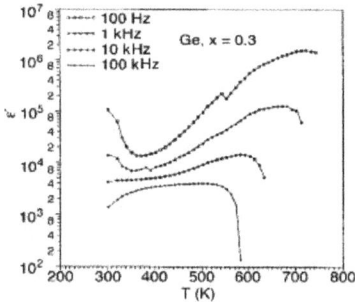
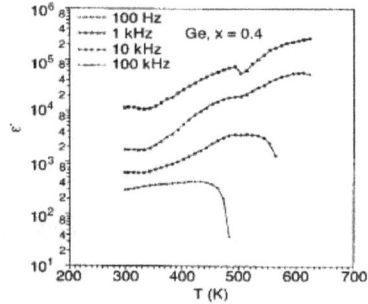

b)

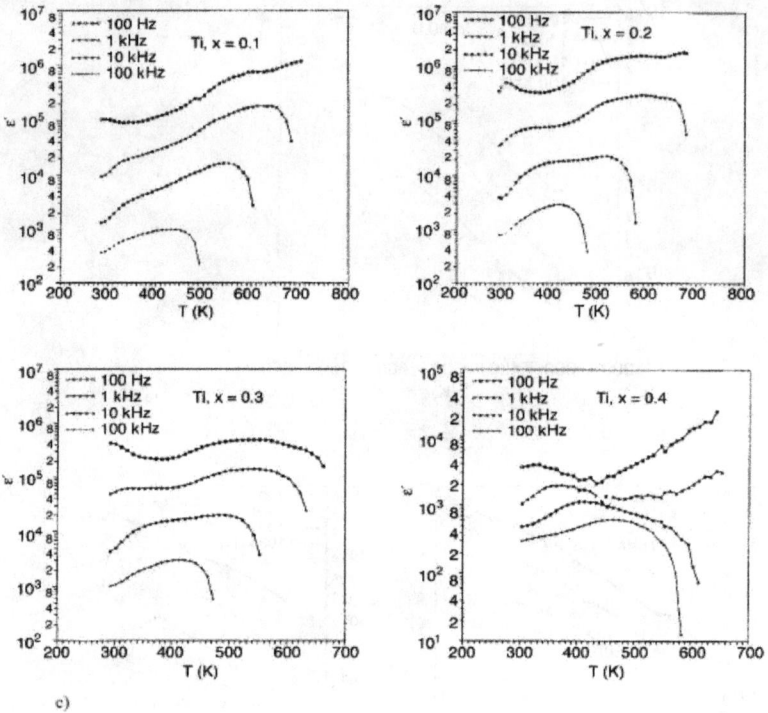

Fig. 11 The variation of the dielectric constant versus temperature at four frequencies for a) Cu ferrite, b) Cu–Ge ferrite, c) Cu–Ti ferrite.

(III) Copper doped magnetite:

AC conductivity

All AC parameters (AC conductivity, dielectric constant and dielectric loss) are studied over the frequencies range from 50 Hz up to 5MHz for $Cu_xFe_{3-x}O_{4+\delta}$ at room temperature. The variations of these parameters are shown in Fig. 12. It is observed from the figures that the conductivity at low frequency is frequency independent (or nearly) for the low copper concentrations and then becomes frequency dependent with the increase of Cu content increases (x > 0.4). By the increase in frequency, the conductivity

Eng. Magnetic, Dielectric and Microwave Properties of Ceramics and Alloys   Materials Research Forum LLC
Materials Research Foundations **57** (2019) 113-148   doi: https://doi.org/10.21741/9781644900390-6

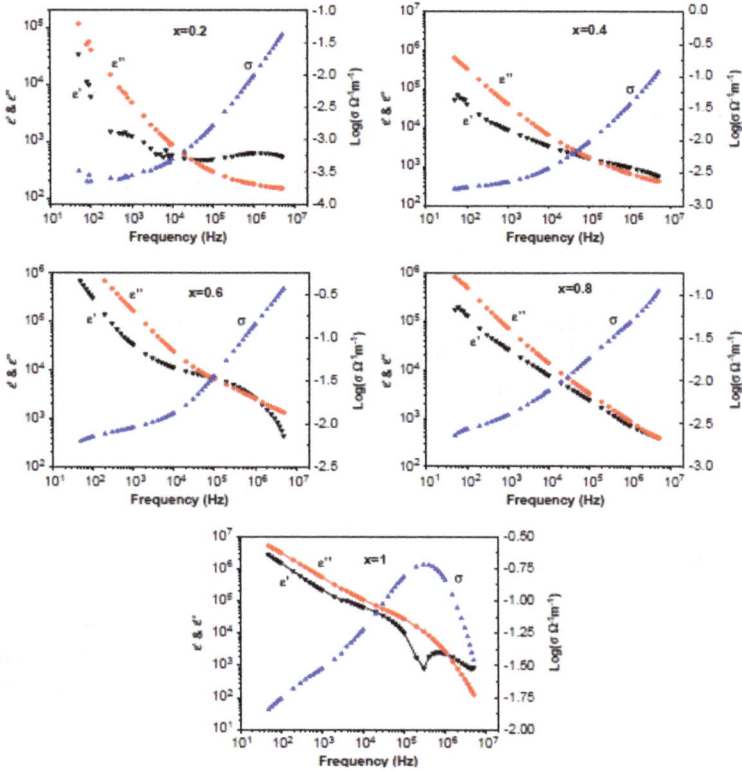

Fig.12. Plots of σ, ε' and ε'' versus frequency at room temperature for x = 0.2, 0.4, 0.6, 0.8 and 1 in $Cu_xFe_{3-x}O_{4+\delta}$

becomes more frequency dependent. But AC conductivity in disordered solids is an increasing function of frequency (any hopping model has this feature) and it is possible to distinguish the different characteristic regions of frequency [27]. In this model, the conductivity is constant at low frequencies where an infinite path was taken in transportation. The conductivity increases strongly with frequency, in a certain region of frequencies where the transport is dominated by contributions from hopping infinite clusters. Finally, the region where the high frequency cutoff starts to play a role (here saturation will be reached) is encountered. Heikes and Johnson explained the electrical conduction mechanism in terms of the electron hopping model [28]. With other words, the conduction mechanism may be related to the electron hopping between two

Eng. Magnetic, Dielectric and Microwave Properties of Ceramics and Alloys   Materials Research Forum LLC
Materials Research Foundations **57** (2019) 113-148   doi: https://doi.org/10.21741/9781644900390-6

octahedral sites in the spinel lattice and a transition between $Fe^{2+} \leftrightarrow Fe^{3+}$ ions or $Cu^{2+} \leftrightarrow Cu^{+}$ might take place [29].

In the range of frequency up to 10^3 Hz samples with a low concentration of x up to 0.4, does not affect the exchange mechanism. But the effect appears above this frequency range. Saturation was not reached for all samples, except the sample of x = 1 ($CuFe_2O_4$) with abnormal behavior. Accordingly, a maximum value for σ was reached at F = 5 x 10^5 Hz followed by a decrease in s with any further increase in frequency up to 5 MHz, such decrease is explained by the obstruction of the hopping mechanism by the applied field. Hence the conductivity will decrease by increasing frequency.

The behaviors of both dielectric constant and dielectric loss against frequency at room temperature are shown in Fig. 12. The general trend for dielectric constant and dielectric loss is a decreasing behavior by the increase of frequency, where the dielectric constant for Cu ferrite reduced from 3 x 10^6 at 50 Hz to 10^3 at 5 MHz. The inhomogeneous structure of these ferrites gives the high values of dielectric constant.

According to Fig. 12, the studied system for both ε' and ε" show a decrease with increasing frequency. However, ε" decreases faster than ε' within the same frequency range. But the values of the dielectric constants (ε') in the high-frequency range become closer to the value of dielectric loss (ε"). All samples exhibit dispersion due to Maxwell interfacial polarization [30], in agreement with the phenomenological theory [9]. This dielectric behavior can be explained using the similarity between the polarization mechanism process and that of the conduction process in ferrite. Iwauchi [23] has pointed out that there is a strong correlation between the conduction mechanism and the dielectric behavior of ferrites. Heikes and Johnson [28] stated that the electronic exchange in such ferrite may be considered as $Fe^2 + Cu^2 \leftrightarrow Fe^{3+} Cu$. It is well known, that the field inside the medium reduces due to the effect of polarization.

Therefore, as the frequency is increased the dielectric constant may decrease. Also, such a decrease may be related to the fact that the electronic exchange between ferrous (Fe^{2+}) and ferric ($Fe3^+$) ions cannot follow the external applied field beyond a certain frequency.

The frequency dependence of the AC conductivity is depicted in Fig. 13 at different temperatures for all samples.

Eng. Magnetic, Dielectric and Microwave Properties of Ceramics and Alloys   Materials Research Forum LLC
Materials Research Foundations **57** (2019) 113-148   doi: https://doi.org/10.21741/9781644900390-6

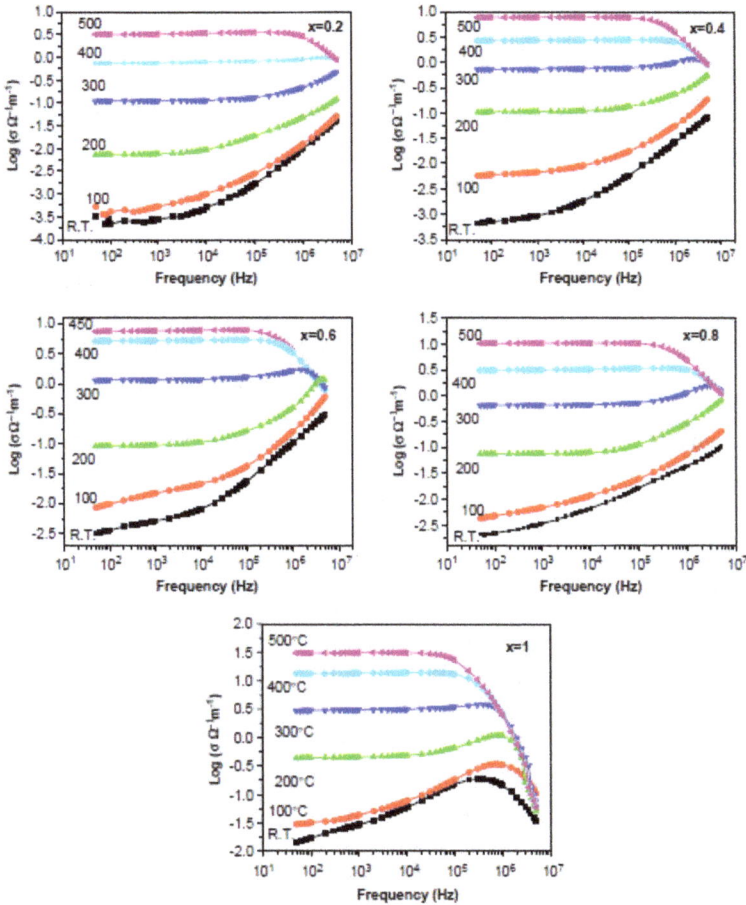

Fig. 13. Variation of AC conductivity with frequency at fixed points of temperature for different Cu concentrations (x) for $Cu_xFe_{3-x}O_{4+\delta}$.

It is clear from the figure that the dispersion in conductivity occurs at low frequencies within the low-temperature range and vice versa. As Cu^{2+} ions replaced Fe^{3+} ions, the dispersion shift towards lower frequencies. This dispersion was explained on the basis of interfacial polarization due to the inhomogeneous structure of ferrite material. Generally, the AC conductivity dispersion decreases with increasing the temperature for all samples. The AC conductivity seems to be independent of frequency. At relatively high temperatures, until certain frequency at which the conductivity begins to decrease, where the applied field frequency obstructs hopping conduction as mentioned before.

4.   Frequency dependence of the dielectric constant (ε') and dielectric loss (ε''):

The variation of the dielectric constant and dielectric loss against the frequency is shown in Figs. 14 and 15 at different temperature for the studied samples. Generally, the values of dielectric constant (ε') are relatively high for the studied ferrite ($Cu_xFe_{3-x}O_{4+\delta}$) and ε' shows inversely peaking behavior with frequency. The minimum value was observed to be depended on temperature and composition. By the increase in temperature, the frequency shifts to lower frequencies as the minimum occurs. But as Cu^{2+} content increasing the minimum appears at relatively lower temperatures. Koop's phenomenological theory and the Rezlescu model have explained the behavior of dielectric constant for the compositions under investigation [9], [31]. Koop's model stated that ferrite samples with homogeneous structure can be imagined as a system consisting of high conductive grains with ε_1, σ_1 and thickness d_1 separated by highly resistive thin grain boundaries with ε_2, σ_2 and thickness d_2.

Rezlescu model may be explained the peaking behavior of ε' with frequency. Using this model, the peaks of dielectric constant (ε') curves may confirm the presence of two different types of charge carriers made a collective contribution to the polarization [31]. For the studied ferrite, the conduction process may be related to the presence of two types of charge carriers, a transfer between Fe^{2+} and Fe^{3+} due to electron transformation (n-type), and between Cu^+ and Cu^{2+} as hole exchange at the octahedral sites (p-type) [32, 33]. These two mechanisms can be represented as: $Fe^{2+} \leftrightarrow Fe^{3+} + e^-$ and $Cu^{2+} \leftrightarrow Cu^+ + e^+$.

Since the displacement direction of holes is opposite to that of electrons under the application of applied field and the mobility of holes is relatively very small with respect to the electrons. So, the polarization result for both types of charge carriers will give peaking behavior as shown in Fig. 14. The peak shift to a higher frequency with the increase in temperature can be related to the corresponding increase in mobility of the charge carriers with temperature. This behavior was also observed for Cu–Zn, Cu–Mn and Cu–Ni ferrite [31], Ni–Zn ferrite [34] and Cu–Cr ferrite [35].

It is observed from Fig. 15 that the dielectric loss (ε'') decreases with the increase in frequency. According to Smith and Wijn [36] for the same temperature, the ratio between the dielectric loss to the AC conductivity is inversely proportional to the applied frequency where,

$$\varepsilon'' = 1.8 \times 10^{10} (\frac{\sigma}{F}) \tag{6}$$

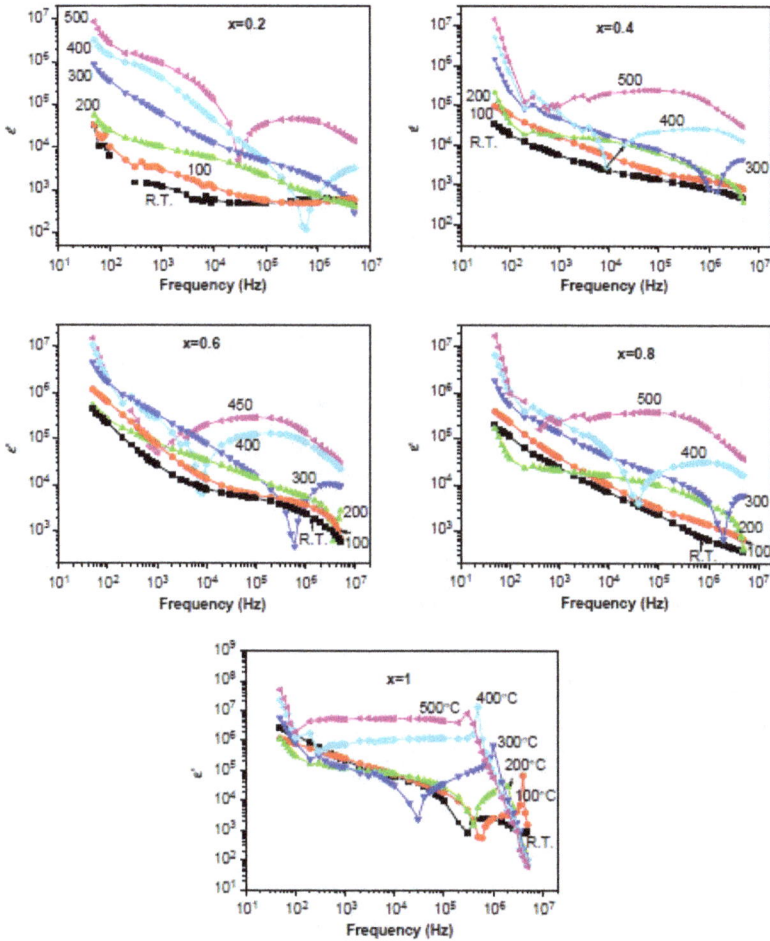

Fig.14. The relation between the dielectric constant (ε'') and frequency at fixed points of temperature for different concentrations (x) in $Cu_xFe_{3-x}O_{4+\delta}$

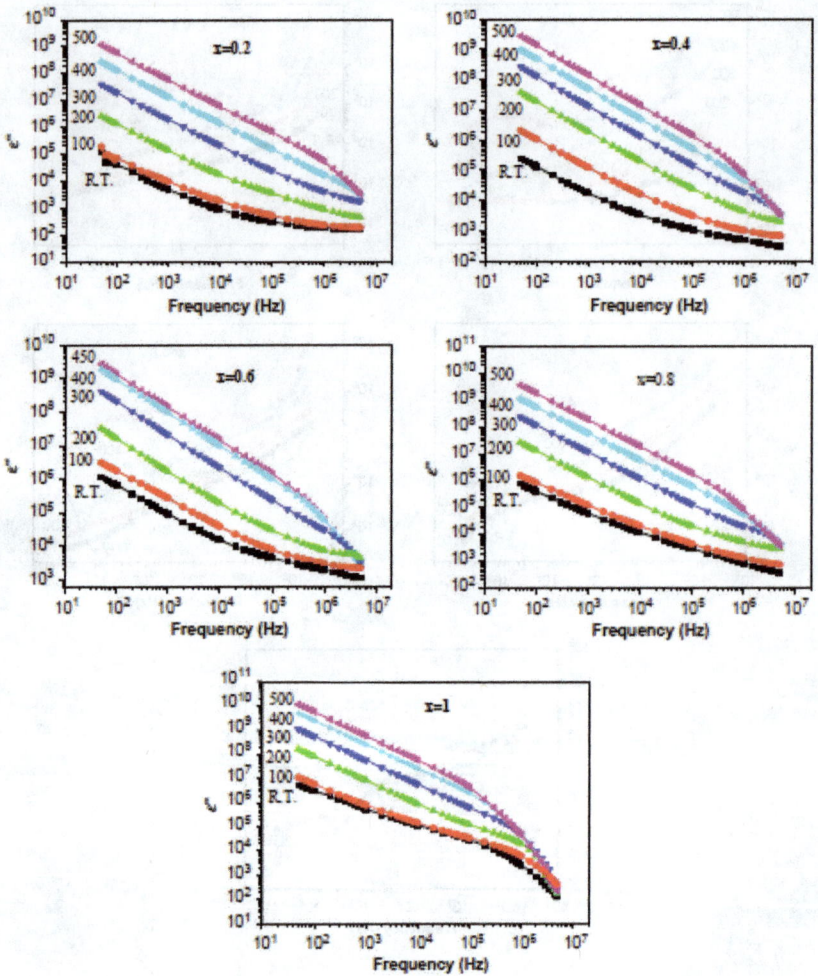

Fig.15. The relation between the dielectric loss (ε'') and frequency at fixed points of temperature for different concentrations (x) in $Cu_xFe_{3-x}O_{4+\delta}$

Eng. Magnetic, Dielectric and Microwave Properties of Ceramics and Alloys   Materials Research Forum LLC
Materials Research Foundations **57** (2019) 113-148   doi: https://doi.org/10.21741/9781644900390-6

5.   Dielectric loss tangent behavior

The variation of dielectric loss tangent (tan δ) against frequency is shown in Fig. 16 at different temperatures for the studied samples. A peaking behavior for the studied samples was observed for that curve which is a function of both composition and temperature. The peak shifted towards lower frequencies with the increase in temperature. As copper ion content increases, new peaks at lower temperatures were observed and the peaks were shifted towards lower frequencies and as the temperature increases the height of the peak increases.

The peaking behavior of tanδ may be explained using the strong correlation between the conduction mechanism and the dielectric behavior in ferrite [31, 37]. Accordingly, when the hopping frequency of the electrons between Fe^{2+} and $Fe^{3}+$ ions is equal to the external applied field a peak is expected, and in this case: $\omega\tau = 1$ and $\tau = 1/2P$ where τ, is the relaxation time of the hopping process, ω is the angular frequency of the external applied field and P is the jumping probability per unit time [38, 39].

The shift of the peak in loss tangent towards low frequency by the increase in temperature indicates that the jumping probability, p, decreases with increasing temperature. Also, the figure shows that the jumping probability decreases as Copper ions content increases which may be ascribed to the decrease of iron ion in the B-site (which is responsible for the polarization in this ferrite). The similar behavior was observed in zinc-substituted magnesium-rich manganese ferrite [40], Nd–Cu ferrite [41], Gd–Cu ferrite [42], Ni–Zn ferrite [34] and also in Cu–Zn ferrite [43].

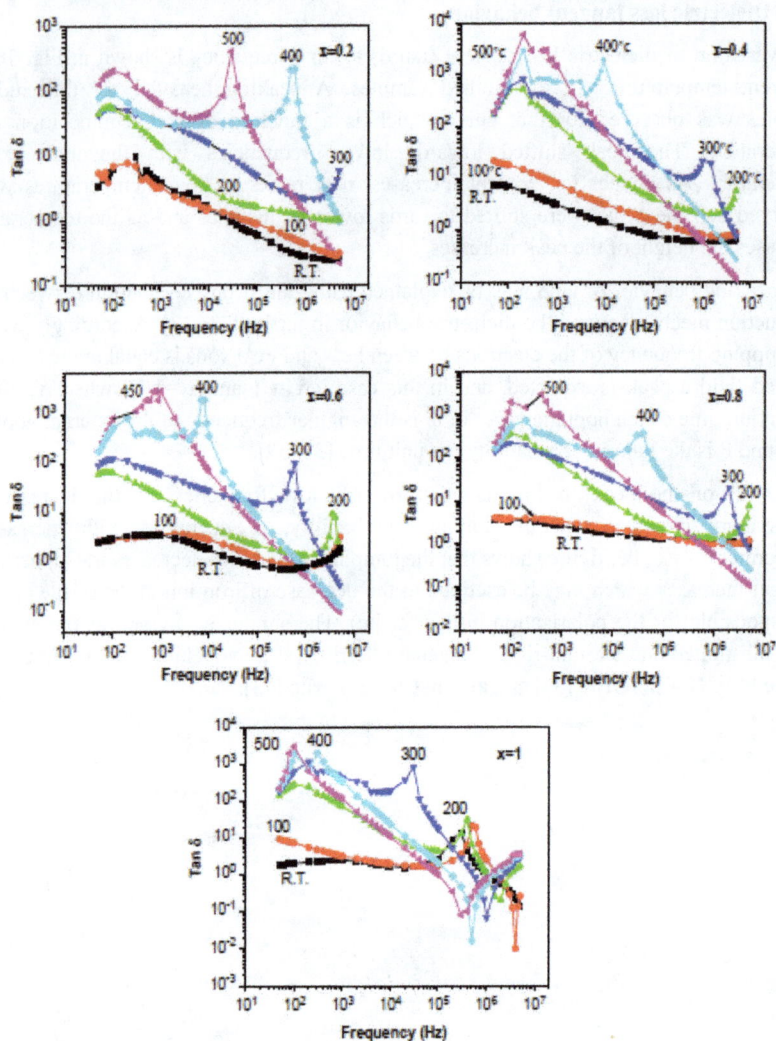

Fig. 16 Variation of the dielectric loss tangent with frequency at different temperatures for samples in $Cu_xFe_{3-x}O_{4+\delta}$

6.   Composition dependence of σ, ε' and ε'':

The dependence of AC conductivity, dielectric constant and dielectric loss on Cu ion content is shown in Fig. 17 at room temperature for 10^2 Hz. The figure shows that all the parameter (σ, ε' and ε'') increase as copper ion substitution increases. As Cu content increases a number of vacancies may exist at the iron site. The thermal dissociation of oxygen may initiate such vacancies, which in turn increases the number of electrons [44]. Also, during the sample preparation a number of ferrous ions may be formed, which in turn can increase the hopping process which may increase in σ, ε' and ε'' as Cu content increases. Similar behavior has also been observed at higher frequencies (10^3–10^5 Hz) for all compositions at room temperature. Also according to Tawfik et al. [45], the formation of vacancies may produce as the substitution of Cu ions. Anion vacancies retard the jump frequency away from the applied field frequency and then the

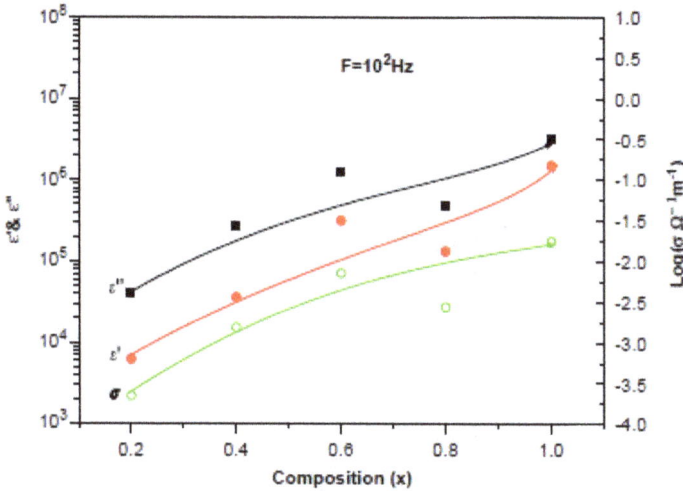

Fig. 17 Composition dependence of AC conductivity, dielectric constant and dielectric loss at room temperature at fixed frequency 10^2 Hz.

dielectric constant will decrease. But in our case of the studied system the increase in ε' suggests that the formed vacancies are cations. Table 4 tabulated the obtained results at different frequencies.

Table 4: Frequency and composition dependence of σ, $\in'$, $\in''$ and tan δ for $Cu_xFe_{3-x}O_{4+\delta}$

Frequency (Hz)	Dielectric symbol	x = 0.2	0.4	0.6	0.8	1.0
10^2	log σ (Ω^{-1} m^{-1})	−3.66	−2.82	−2.15	−2.57	−1.76
	ε'	$6.19e^3$	$3.57e^4$	$3.12e^5$	$1.33e^5$	$1.5e^6$
	ε''	$3.97e^4$	$2.76e^5$	$1.27e^6$	$4.87e^5$	$3.16e^6$
	tan δ	6.41	7.72	4.07	3.68	2.04
10^3	log σ (Ω^{-1} m^{-1})	−3.57	−2.72	−2.042	−2.4	−1.52
	ε'	$1.24e^3$	$8.61e^3$	$3.3e^4$	$2.65e^4$	$2.22e^5$
	ε''	$4.86e^3$	$3.4e^4$	$1.63e^5$	$7.19e^4$	$5.4e^5$
	tan δ	3.93	3.96	4.96	2.72	2.44
10^4	log σ (Ω^{-1} m^{-1})	−3.3	−2.49	−1.88	−2.12	−1.22
	ε'	$5.82e^2$	$3.3e^3$	$1.13e^4$	$7.7e^3$	$6.41e^4$
	ε''	$9.03e^2$	$5.9e^3$	$2.4e^4$	$1.37e^4$	$1.08e^5$
	tan δ	1.55	1.76	2.13	1.78	1.68
10^5	log σ (Ω^{-1} m^{-1})	−2.57	−1.46	−0.85	−1.32	−0.84
	ε'	$5.29e^2$	$6.24e^2$	$2.57e^3$	$7.07e^2$	$2.38e^3$
	ε''	$2.42e^2$	$8.81e^2$	$2.61e^3$	$8.58e^2$	$2.6e^3$
	tan δ	0.46	0.71	0.99	1.22	1.09

7. Determination of the frequency exponential factor (S)

Jonscher [46] stated that the real part of AC conductivity may be written as:

$$\sigma = \sigma_{DC} + \sigma_{AC} = \sigma_{DC} + \omega^S \qquad (7)$$

The first part of the equation represents the DC conductivity and the second one is the pure AC conductivity; where A has the conductivity unit, S is a dimensionless parameter and the angular frequency (ω) was measured as conductivity σ. As mentioned before S values are usually between 0.4 and 0.8 [47]. Over the studied range of frequencies, S was evaluated for all the samples at room temperature and the results were explained owing to Pike model [2]. The values of S in the present work were calculated and presented in Fig. 18. The results agree with the results of some other workers [48-50].

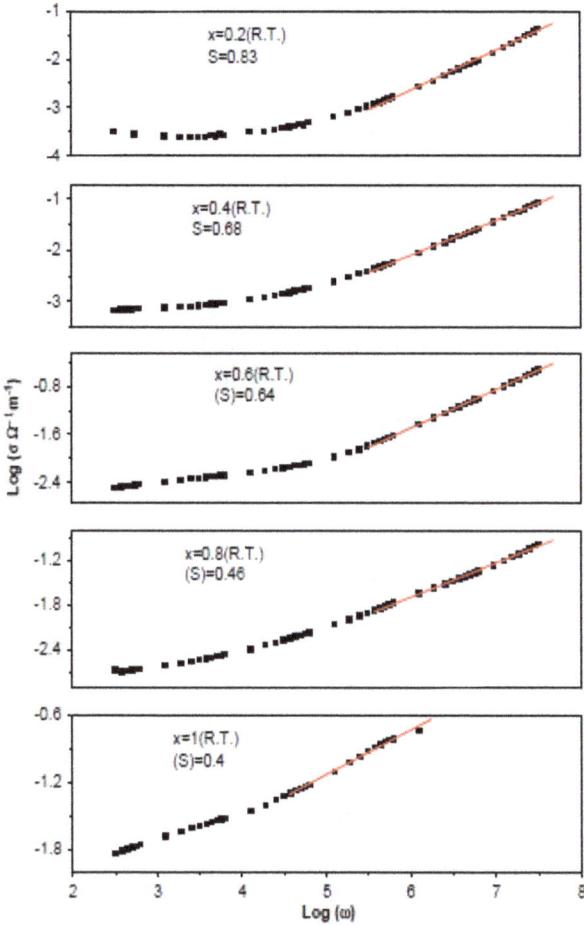

Fig. 18 The variation of the exponential factor (S) with angular frequency (log ω) at room temperature for all samples.

Conclusions

I. It observes from the AC conductivity relation for $Li_{0.5+0.5x}Ge_xFe_{2.5-1.5x}$ (x= 0.0, 0.2, 0.3 and 0.5) ferrite that it is independent on frequencies at the lower stage and, above a certain frequency, it almost dependent, by increases value with the increase in frequency. the activation energy tends to decrease by the increase in frequency but the conductivity increases at all temperatures by the increase in frequency, and a marked change is prominent at lower temperatures which is explained by Pike. The maximum $\tan\delta$ reflects the rapid decrease in ε' as the increase in frequency due to the dispersion and slowly decreases in ε''.

II. The AC conductivity exhibits dispersion with respect to the frequency and the hopping model has been used to interpret the conduction mechanism. According to the Maxwell–Wagner model, the hopping rate ($g = \tau^{-1}$) was found to be about 2×10^6 s^{-1} and 5×10^5 s^{-1} for Cu–Ge and Cu–Ti ferrites, respectively. The average value of the activation energy of conduction mechanism of Cu–Ti (0.255 eV) seems to be lower than that of Cu–Ge (0.33 eV) and this may be due to the screening role and distribution of each tetravalent ion.

III. The dielectric behavior can be explained in terms of the electron exchange between Fe^{2+} and Fe^{3+}, and the hopping of a hole between Cu^{2+} and Cu^+ ions at B-sites, suggesting that the polarization in these compositions is similar to that of the conduction process in ferrites. Abnormal behavior (peaks) was observed in $\tan\delta$ curves at relatively high temperatures. Such relaxation peaks take place when the jumping frequency of localized electrons between Fe^{2+} and Fe^{3+} ion equals that of the applied AC electric field. The broad maxima peaks tend to shift towards lower frequency as the temperatures increases. The exponent factor (S) was calculated and found to be acceptable in the range of the reported values.

References

[1] A.M.M. Farea, Shalendra Kumar, Ali Yousef, Chan Gyu Le and Alimuddin, Structure and electrical properties of $Co_{0.5}Cd_xFe_{2.5-x}O_4$ ferrites, Journal of Alloys and Compounds, 464 (2008) 361. https://doi.org/10.1016/j.jallcom.2007.09.126

[2] Y. Köseoğlu, M. Bay, M. Tan, A. Baykal, H. Sözeri, R. Topkaya and N. Akdoğan, Magnetic and dielectric properties of $Mn_{0.2}Ni_{0.8}Fe_2O_4$ nanoparticles synthesized by PEG-assisted hydrothermal method, Journal of Nanoparticle Research, 13 (2011) 2235. https://doi.org/10.1007/s11051-010-9982-6

[3] Razia Nongjai, Shakeel Khan, K. Asokan, Hilal Ahmed, and Imran Khan, Magnetic and electrical properties of In doped cobalt ferrite nanoparticles, Journal of Applied Physics 112, (2012) 084321. https://doi.org/10.1063/1.4759436

[4] S. Mahalakshmi, K. SrinivasaManja, and S. Nithiyanantham, Electrical Properties of Nanophase Ferrites Doped with Rare Earth Ions, Journal of Superconductivity and Novel Magnetism, 27 (2014) 2083. https://doi.org/10.1007/s10948-014-2551-y

[5] Baykal, Al., Kasapoglu, N., Koseoglu, Y.K., Toprak, M.S., and Bayrakdar, H., CTAB-assisted hydrothermal synthesis of $NiFe_2O_4$ and its magnetic characterization. J. Alloys. Compd. 464 (1-2) (2008) 514–518. https://doi.org/10.1016/j.jallcom.2007.10.041

[6] S. U. Haque, K. K. Saikia, G. Murugesan, and S. Kalainathan, "A study on dielectric and magnetic properties of lanthanum substituted cobalt ferrite," J. Alloys Compounds, 701 (2017) 612–618. https://doi.org/10.1016/j.jallcom.2016.11.309

[7] M. Pita et al., Synthesis of cobalt ferrite core/metallic shell nanoparticles for the development of a specific PNA/DNA biosensor, J. Colloid Interface Sci., 321(2) (2008) 484–492. https://doi.org/10.1016/j.jcis.2008.02.010

[8] A. K. Nikumbh et al., Structural, electrical, magnetic and dielectric properties of rare-earth substituted cobalt ferrites nanoparticles synthesized by the co-precipitation method, J. Magn. Magn. Mater., 355 (2014) 201–209. https://doi.org/10.1016/j.jmmm.2013.11.052

[9] C. G. Koops, On the Dispersion of Resistivity and Dielectric Constant of Some Semiconductors at Audio frequencies, Phys. Rev., 83 (1951)121. https://doi.org/10.1103/PhysRev.83.121

[10] G. E. Pike, AC Conductivity of Scandium Oxide and a New Hopping Model for Conductivity, Phys. Rev. B, 6 (1972) 1572. https://doi.org/10.1103/PhysRevB.6.1572

[11] M. Pollak, On the frequency dependence of conductivity in amorphous solids, Phil. Mag., 23 (1971) 519. https://doi.org/10.1080/14786437108216402

[12] S. R. Elliot, A theory of a.c. conduction in chalcogenide glasses, Phil. Mag., 36 (1977) 1291. https://doi.org/10.1080/14786437708238517

[13] S. A. Mazen, M. H. Abdallah, M. A. El-Ghandoor, and H. A. Hashem, Dielectric behaviour of $Cu_{1-x}Ti_xFe_2O_4$ ferrites, phys. stat. sol., 144 (1994) 461. https://doi.org/10.1002/pssa.2211440227

[14] J. Volger, Dielectric properties of solids in relation to imperfections, Prog. Semicond., 4 (1960) 207

[15] F. Haberey and H. P. Wijn, Effect of temperature on the dielectric relaxation in polycrystalline ferrites, phys. stat. sol.(a), 26 (1968) 231. https://doi.org/10.1002/pssb.19680260124

[16] N. F. Mott and E. A. Davis, Electronic Processes in Non-crystalline Material, Oxford Press, UK, (1979)

[17] N. F. Mott, Electrons in disordered structures, Adv. Phys., 16 (1967) 49. https://doi.org/10.1080/00018736700101265

[18] J. C. Dyre, The random free energy barrier model for ac conduction in disordered solids, J. Appl. Phys., 63 (1988) 2456. https://doi.org/10.1063/1.341681

[19] J. Friedel, On some electrical and magnetic properties of metallic solid solutions, Can. J. Phys., 34 (1956) 1190. https://doi.org/10.1139/p56-134

[20] S. A. Mazen, F. Metawe, and S. F. Mansour, IR absorption and dielectric properties of Li-Ti ferrite, J. Phys. D: Appl. Phys., 30 (1997) 1799. https://doi.org/10.1088/0022-3727/30/12/018

[21] N. Rezlescu and E. Rezlescu, Dielectric properties of copper containing ferrites, phys. stat. sol. (a), 23 (1974) 575. https://doi.org/10.1002/pssa.2210230229

[22] S. A. Mazen and H. M. Zaki, Ti^{4+} and Ge^{4+} ionic substitution in Cu-ferrite, electrical conductivity and thermoelectric power, J. Magn. Magn. Mater., 248 (2002) 200. https://doi.org/10.1016/S0304-8853(02)00281-0

[23] K. Iwavachi, IR absorption and dielectric properties of Li-Ti ferrite, J. Appl. Phys., 10 (1971) 1520

[24] N. Rezlescu and E. Cuciureanu, Cation distribution and curie temperature in the copper-manganese-zinc ferrites, J. Phys. Chem. Solids, 32 (1971) 1096. https://doi.org/10.1016/S0022-3697(71)80356-6

[25] N. Rezlescu and E. Cuciureanu, Cation distribution and curie temperature in some ferrites containing copper and manganese, phys. stat. sol. (a), 3 (1970) 573. https://doi.org/10.1002/pssa.19700030403

[26] B. Tareev, Physics of Dielectric Materials, Mir Pub., Moscow, (1975)

[27] H. Bottger, V.V. Bryksin, Hopping Conduction in Solids, Berlin, (1985)

[28] R.R. Heikes, and W.D. Johnson, Mechanism of Conduction in Li Substituted Transition Metal Oxides, J. Chem. Phys., 26 (1957) 582. https://doi.org/10.1063/1.1743350

[29] J.H. Jonker, Analysis of the semiconducting properties of cobalt ferrite, J. Phys. Chem. Solids, 9 (1959) 165. https://doi.org/10.1016/0022-3697(59)90206-9

[30] J.C. Maxwell, Electricity and Magnetism, Oxford Press, London, 1 (1973)

[31] N. Rezlescu and E. Rezlescu, Abnormal dielectric behaviour of copper containing ferrites, Solid State Commun., 14 (1974) 69. https://doi.org/10.1016/0038-1098(74)90234-8

[32] X.-X Tang, A. Manthiram and J.B. Goodenough, Copper ferrite revisited, J. Solid State Chem., 79 (1989) 250. https://doi.org/10.1016/0022-4596(89)90272-7

[33] B.L. Patil, S.R. Sawant and S.A. Patil, Temperature Dependence of Electrical Resistivity and Thermoelectric Power in Cu Ti Fe O Ferrites, Phys. Status Solidi (a), 133 (1992) 147. https://doi.org/10.1002/pssa.2211330115

[34] R.V. Mangalaraja, S. Ananthakumar, P. Manohar and F.D. Gnanam, Magnetic, electrical and dielectric behaviour of $Ni_{0.8}Zn_{0.2}Fe_2O_4$ prepared through flash combustion technique, J. Magn. Magn. Mater., 253 (2002) 56. https://doi.org/10.1016/S0304-8853(02)00413-4

[35] M.A. El Hiti, M.A. Ahmed, M.M. Mossad and S.M. Ahia, Dielectric behaviour of Cu-Cr ferrites, J. Magn. Magn. Mater., 150 (1995) 399. https://doi.org/10.1016/0304-8853(95)00281-2

[36] J. Smith, H.P. Wijn, Ferrites, Wiley, New York, (1959)

[37] K. Iwauchi, Dielectric properties of fine particles of Fe3O4 and some ferrites, J. Appl. Phys., 10 (1971) 520. https://doi.org/10.1143/JJAP.10.1520

[38] N. Nanba, Distribution of cation vacancies in copper ferrites with a stoichiometric excess of oxygen, J. Appl. Phys., 53 (1982) 695. https://doi.org/10.1063/1.329978

[39] M.B. Reddy, and P.V. Reddy, Low-frequency dielectric behaviour of mixed Li-Ti ferrites, Physica D, 24 (1991) 975. https://doi.org/10.1088/0022-3727/24/6/025

[40] K.P. Thummer, H.H. Joshi and R.G. Kulkarni, Electrical and dielectric properties of zinc substituted magnesium rich manganese ferrites, J. Mater. Sci. Lett., 18 (1999) 1529. https://doi.org/10.1023/A:1006654720054

[41] J.W. Chen, J.C.Wang and Y.F. Chen, Study of dielectric relaxation behavior in Nd_2CuO_4, Physica C, 289 (1997) 131. https://doi.org/10.1016/S0921-4534(97)01577-3

[42] J.B. Shi, Y. Hsu and C.L. Lin, Dielectric properties of Gd_2CuO_4, Physica C, 299 (1998) 272. https://doi.org/10.1016/S0921-4534(98)00038-0

[43] A.A. Sahar and S.A. Rahman, Dielectric properties of rare earth substituted Cu–Zn ferrites, Phys. Status Solidi (a), 200 (2003) 415. https://doi.org/10.1002/pssa.200306663

[44] Y. Yamamoto, A. Makino and T. Nikaidou, Low loss of fine grained Mn-Zn ferrite, J. Phys. IV, Collog (France), 7 (1997) c1-121. https://doi.org/10.1051/jp4:1997139

[45] A. Tawfik and O.M. Hemeda, Effect of vacancy jump rate on the permeability and dielectric properties of $Ni_{0.65}Zn_{0.35}Cu_xFe_{2-x}O_4$, Mater. Lett., 56 (2002) 665. https://doi.org/10.1016/S0167-577X(02)00573-6

[46] A. Jonscher, Dielectric Relaxation in Solids, Chelsea Dielectric Press, London, (1983)

Eng. Magnetic, Dielectric and Microwave Properties of Ceramics and Alloys   Materials Research Forum LLC
Materials Research Foundations **57** (2019) 113-148   doi: https://doi.org/10.21741/9781644900390-6

[47] N.F. Mott and E.A. Davis, Electronic Processes in Non-Crystalline Materials, second ed., Clarendon Press, Oxford, (1979)

[48] S.A. Mazen, Infrared absorption and dielectric properties of Li–Cu ferrite, Mater. Chem. Phys., 62 (2000) 139. https://doi.org/10.1016/S0254-0584(99)00158-3

[49] S.A. Mazen and H.M. Zaki, AC conductivity of Li-Ge ferrite, J. Phys. D, 28 (1995) 609. https://doi.org/10.1088/0022-3727/28/4/002

[50] S.A. Mazen and H.A. Dawoud, Temperature and composition dependence of dielectric properties in Li–Cu ferrite, Mater. Chem. Phys., 82 (2003) 557. https://doi.org/10.1016/S0254-0584(03)00200-1

Eng. Magnetic, Dielectric and Microwave Properties of Ceramics and Alloys   Materials Research Forum LLC
Materials Research Foundations **57** (2019) 149-174   doi: https://doi.org/10.21741/9781644900390-7

Chapter 7

Structure–Property Relations in Rare-Earth Doped Manganite Perovskites

K. Sakthipandi[1*], Aslam Hossain[2], G. Rajkumar[3]

[1]Department of Physics, Sethu Institute of Technology, Kariapatti - 626 115, Tamil Nadu, India

[2]Department of Physical and Inorganic Chemistry, Institute of Natural Science, Ural Federal University Yekaterinburg, Russia

[3]Department of Physics, Easwari Engineering College, Chennai - 600 089, Tamil Nadu, India

sakthipandi@gmail.com

Abstract

The perovskite with $R_{1-x}A_xMnO_3$ (R = rare-earth and A = divalent metals) formula has worldwide attention due to its fascinating electrical and magnetic properties. A-site doped manganite perovskite oxides were broadly investigated based on theoretical and experimental point-of-view due to interesting phenomenon such as charge ordering, orbital ordering, colossal magnetoresistance, order-disorder transition, phase separation scenario, etc. Such kind of manganite system has two types of transitions as an insulator to metal transition and paramagnetic to ferromagnetic transition. The significance of perovskite materials and their electronic, magnetic or structural phase transition temperatures were analyzed with different experimental results.

Keywords

Perovskites Oxide, Magnetic Properties, Ionic Radius, Colossal Magnetoresistance, Curie Temperature

Contents

Eng. Magnetic, Dielectric and Microwave Properties of Ceramics and Alloys   Materials Research Forum LLC
Materials Research Foundations **57** (2019) 149-174   doi: https://doi.org/10.21741/9781644900390-7

1.   Introduction

Materials play an important role in the progress of human civilization. Historical evidence explores the step-wise development of life according to the use of materials such as stone in stone-age, bronze in bronze-age, etc. The gradual development in the search for new materials has made a strong important impact on human activities over thousands of years. Materials science is an interdisciplinary branch of science which deals with the elements of applied chemistry and physics as well as mechanical, electrical, civil and chemical engineering. The research on properties of matter and its applications in different field of science and engineering is an interesting journey.

In recent years, materials science is fast moving to the forefront thanks to the developments achieved in nanoscience and nanotechnology. The aim of materials science is to study and understand the fundamentals of materials, so that new materials with interesting properties might be discovered for different engineering and biomedical applications. The study of physicochemical properties as a function of its structure and origin of such property changes are the well-known thought in material science. Further,

Eng. Magnetic, Dielectric and Microwave Properties of Ceramics and Alloys   Materials Research Forum LLC
Materials Research Foundations **57** (2019) 149-174   doi: https://doi.org/10.21741/9781644900390-7

it supports to develop new and improved materials to meet the requirement in various fields and device processes for manufacturing such materials.

The subject deals with the study of a wide range of materials with technological and biological importance. The modern world is heavily thrust on novel and smart materials for telecommunication transport, biomedical, energy storage applications [1-2].

1.1   Classification of Materials

Materials can be classified as metals, composites, semiconductors, ceramics, polymers, ferrites, perovskites and biomaterials.

Metals: Metals have high thermal and electrical conductivity. Metals are strong but undergo deformation when mechanical forces are applied. Generally, the metallic bond was found to bound the atoms together in a metal system. This type of bond is possible when valence electron separates from atoms and spread throughout the Electron Sea and created forces to bind ions together. Usually, pure metals are poor in mechanical strength and not stable in different active environments. The addition of other suitable metal to make an alloy helps to improve the desired qualities. (Examples: brass: (Copper 95 % + Zinc 5%) bronze: (Copper (88%) + Tin (12%).

Ceramics: These are the kinds of inorganic material such as oxides, carbides, silicates of metals, nitrides, etc. These types of materials are neither full crystalline nor amorphous. The atoms present in ceramics are either positive or negative which are bound by very strong Coulomb forces. These ceramic materials are generally studied by high strength under compression, low ductility; usually insulators to heat and electricity. Examples: glass, porcelain and many minerals.

Composite materials: The artificial combination of different phase content materials is called composite materials. The properties of these materials are different than where individual materials are used. Such kind of materials is classified by their matrix materials. Different types of main composites are polymer-matrix, metal-matrix, and ceramic-matrix. The properties of composite materials strongly depend on minute proportions of impurities to enhance the electrical conductivity. Doping of III/IV group element with silicon (Si), germanium (Ge) and gallium arsenide (GaAs) are notable examples for composites on semi-conducting materials.

Advanced Materials: These materials are used in intricate and sophisticated High-Tech devices such as electronic gadgets, computers, air/space-crafts, etc. These materials are relatively expensive which are used to either enhance the properties of traditional materials or to develop high-performance materials with capabilities. Different type of modern application of these materials are in LCDs, optical fibers, integrated circuits, lasers, thermal protection for space shuttle [3], etc.

1.2   Perovskites

$R_{1-x}A_xMnO_3$ (R = rare-earth metal and A = divalent metal) perovskite materials drive their extraordinary properties from a peculiarly strong interaction among the magnetic ordering, electronic transport and structure [4-6]. The behaviour of the colossal magnetoresistive (CMR) is determined by the chemical composition, crystal structure and impurities. Therefore, complete knowledge about the composition and structure affected transport and magnetic properties guides the search for newer manganite materials [5-6]. The magnetic and electronic properties of these perovskites are strongly influenced by their chemical composition and crystallinity. The CMR behaviour is exhibited in three important structures of the perovskite manganite materials such as layered, Ruddlesden-Popper and pyrochlore structure [6]. Therefore, it is essential to know how the structure and chemical composition have been used to gain into the physics of CMR materials.

The existence of CMR behavior in $(La,Pb)MnO_3$ perovskite manganite materials was first discovered in 1969. This perovskite structure has 3 dimensional vortex-sharing MnO_6 octahedral and interstitial divalent ions. The substitution of the rare earth element in $R_{1-x}A_xMnO_3$ perovskite materials controls the oxygen stoichiometry and the ration of Mn^{3+} ions to Mn^{4+} ions. Therefore, the rare earth substitution can cause structural modifications to the cubic structure through the distorted MnO_6 octahedral. The turning of the magnitude of MnO_6 distortion leads the systematic investigations to obtain the desired structural, magnetic and electronic degrees of freedom. This perovskite structure is often distorted strongly by the effect of size mismatch of the cation (tolerance factor) and electronic instabilities (Jahn-Teller distortion). Hence, a review on structure–property relations in rare-earth doped manganite perovskites gives a more in-depth discussion about orbital, spin and charge ordering.

The research with perovskite structure is still motivating to the material scientist due to its numerous industrial as well as technological applications. The number of perovskite structure content materials are huge and some of the members are recognized recently [4]. The general formula of perovskite is ABX_3 (X = oxide, halide, nitride, sulfide) where A is a large metal cation close-packed in layers with oxygen ions and smaller metal ion situated in an octahedrally coordinated hole between the close-packed layers. The presence of small transition metal ion in A-site is the interesting observation in recent years [5]. Generally, after findings of high dielectric loss factor of ABF_3 materials, researchers have started to give attention to ABO_3 type materials only. Currently, hybrid halide perovskite is rocking the market in solar cell and optoelectronic applications.

Eng. Magnetic, Dielectric and Microwave Properties of Ceramics and Alloys   Materials Research Forum LLC
Materials Research Foundations **57** (2019) 149-174   doi: https://doi.org/10.21741/9781644900390-7

Some extensive research works have been done in the last few years with hybrid perovskite and it has crossed a long path overcoming its barriers of limitations.

In the year 1839, German chemist and mineralogist Gustav Rose first time exposed $CaTiO_3$. He has given the name of the mineral $CaTiO_3$ after Lev Alexeievitch Perovsky, a Russian military official and dignitary. Generally, the ideal perovskite-type structure is considered as cubic symmetry with space group *Pm3m*. In cubic perovskite, the B cation is 6-fold coordinated and the A cation is 12-fold coordinated with the oxygen anions. (Ba, Sr)TiO_3 and $CaTiO_3$ show cubic and orthorhombic crystal structure due to the larger size of Ba and Sr than Ca (**Fig. 1**). This smaller size of the Ca^{2+} cations is the reason for the crystal structure of the $CaTiO_3$ prototype which is slightly distorted and thus the cubic symmetry is reduced to orthorhombic.

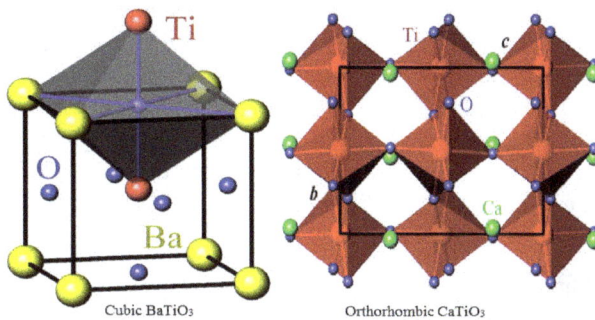

Fig. 1 Simple cubic (left) and orthorhombic crystal structure of perovskite oxide.

In $LaMnO_3$ perovskite compound, the 3d transition Mn atoms are coordinated by six oxygen atoms. Hence it is sufficient to discuss only the octahedral symmetry of the crystal field caused by the oxygen atoms. Out of five 3d orbitals of manganese atoms two orbitals are called e_g orbitals denoted as $3d_{x^2-y^2}$ and $3d_{z^2-r^2}$ point towards oxygen. These e_g orbitals have higher energy than the other three orbitals called t_{2g} orbitals denoted by $3d_{xy}$, $3d_{xz}$ and $3d_{yz}$. When these orbitals are filled with electrons, there are different spin states of Mn ions corresponding to different valence states (2^+, 3^+ and 4^+).

Due to the interplay of structural, magnetic, and transport properties, the perovskites are also called as "Functional Materials". Perovskites exhibit novel phenomena like CMR, metal-insulator (MI) phase transition, Jahn-Teller distortion which attracts significant interest due to their versatile applications in frequency switching devices, magnetic

Eng. Magnetic, Dielectric and Microwave Properties of Ceramics and Alloys   Materials Research Forum LLC
Materials Research Foundations **57** (2019) 149-174   doi: https://doi.org/10.21741/9781644900390-7

refrigeration, sensor technology, solar cells, etc. Hence advanced composite materials can be developed using perovskite materials by doping necessary elements to suit the industrial requirement.

1.3.1 Colossal magnetoresistance

Colossal magnetoresistance (CMR) effect is an interesting physical phenomenon exhibited by perovskite compounds. It attracts greater attention from researchers due to its application in sensor technology and magnetic storage devices. CMR is the very large magnetoresistance associated with magnetic phase transition largely seen in perovskite manganites. It was first reported in 1994 when La-Ca-Mn-O films were investigated. The mechanism in CMR is related to the double exchange (DE) and super exchange (SE) interactions. In DE interaction the electrons move via the intermediate oxygen atoms in the Mn^{3+}-O-Mn^{4+} chain. Here the electron need not have to change its spin direction to move forward. The DE theory proposed by C. Zener gave a qualitative explanation for CMR mechanism [7]. However, according to A.J. Millis et al. DE alone could not explain all aspects of resistivity. In addition to that polaron effects caused by strong electron-phonon from Jahn-Teller splitting of Mn^{3+} ion also plays a vital role in CMR [8]. In SE interaction, the two electrons of the oxygen atom form a strong AFM or FM coupling with the Mn ions. For Mn^{4+}-O-Mn^{4+} the interaction is AFM and that of Mn^{3+}-O-Mn^{4+} may be either AFM or FM.

1.3.2 Jahn-Teller effect

Jahn-Teller distortion is a phenomenon that plays a vital role in the magnetic properties of perovskite materials [9]. According to Jahn-Teller theorem, the systems with orbital energy ground levels are usually unstable and there is a tendency to lower the symmetry and remove the degeneracy. Hence the systems exhibit spontaneous distortion of the octahedron and a lowering of crystal field symmetry accompanied by the splitting of energy levels. This effect is called the Jahn-Teller effect and the distortion produced is called Jahn-Teller distortion. The schematic diagram of the Jahn-Teller effect is shown in Fig. 2.

In manganese oxide MnO_3, the Mn^{3+} ion is surrounded by six oxygen ions giving rise to an octahedron. The electron configuration of Mn^{3+} is $3d^4$. The $3d_{x^2-y^2}$ and $3d_{3z^2-r^2}$ orbitals are oriented towards surrounding oxygen ions. d_{xy} orbital lies around the xy plane while the $3d_{3z^2-r^2}$ orbital is aligned along the z-axis. If this orbital ($3d_{3z^2-r^2}$) is unoccupied then the Mn nucleus is not shielded from the ions along the z axis. Hence the oxygen octahedron is distorted due to electrostatic interaction and compression is caused along the z-axis. On the other hand, if $3d_{x2-y2}$ orbit is unoccupied then an elongation is caused along the z-axis. Thus Jahn-Teller distortion produces e_g orbitals having different energies.

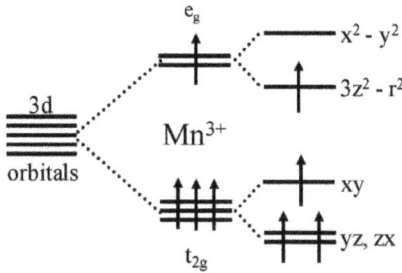

e_g

$x^2 - y^2$

$3z^2 - r^2$

3d

Mn^{3+}

orbitals

xy

yz, zx

t_{2g}

Fig. 2 Splitting of 3d orbitals energy levels - Representation of Jahn-Teller effect.

1.3.3 Doped perovskites

It is evident in so many cases that doping with some foreign materials changes the physical properties such as electrical, chemical, elastic and magnetic properties of the material to a great extent. In some cases, even a slight atomic alteration can have a profound effect on something like electrical conductivity. In such a way certain materials doped in the A or B site of a perovskite material ABO_3. In $R_{1-x}A_xMnO_3$ type oxides, (R = La, Sm, Pr, Nd etc and A = Ca, Na Ba etc) the oxygen has O^{-2} and Mn has Mn^{3+} and Mn^{4+} respectively oxidation state. The substitution of divalent materials promotes the valence of Mn ions from 3 to 4. The charge and spin exchanges between the Mn^{3+} -O^{2-}-Mn^{4+} double exchange (DE) interaction is the primary cause of the exciting magnetic and electrical properties of these materials.

1.3.4 Double exchange interaction

Double exchange (DE) interaction is a magnetic coupling of two cations via oxygen orbital. This concept was used to explain the CMR that occurs at the phase transition temperature by Zener. In the DE process, one of the electrons in the 3d shell of Mn hops to 2p oxygen orbital and simultaneously one electron from the 2p oxygen orbital hops to the 3d shell of another Mn ion. The schematic diagram of DE interaction is shown in Fig. 3. In DE interaction, the spin sign is preserved. Thus the DE leads to FM order of Mn cations. These hoping electrons are responsible for the metallic properties in the compound material. The mean time of hoping is found to be about 10 ps only.

Double exchange

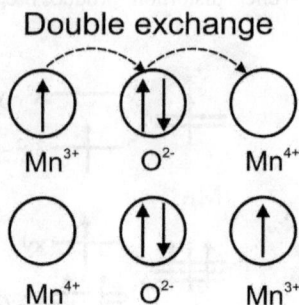

Fig. 3 Schematic representation of DE interaction.

The conditions for DE interaction are stated as follows: One electron should have a non-zero probability to occur in 3d orbitals of both transition metal (Mn) cations. At this juncture, Hund's rule must be fulfilled. Hund's rule states that while filling the molecular orbitals of equal energy, the pairing of electrons does not take place until all such molecular orbitals are singly filled with electrons having parallel spins. Further, the interaction takes place when Mn^{3+}-O-Mn^{4+} bonding angle is close to 180°. The effective hopping of t_{eff} of an electron jumping between two nearest neighbor manganese ions is given by the formula $t_{eff} = t \cos(\theta/2)$ where θ is the angle between t_{2g} spins located at the two sites involved in the transfer of electrons.

A-site substitution of divalent Sr on $LaMnO_3$ promote the number of holes from Mn^{3+} to Mn^{4+} which influence the transition from antiferromagnetic (AFM) insulator state to FM metallic state. The CMR effect evaluated near the transition temperature from FM to PM phase of Sr doped $LaMnO_3$ which is due to the promotion of oxidation state of manganese from Mn^{3+} to Mn^{4+}. Hence DE interaction theory alone cannot explain the behaviour of magnetic and electronic phase diagrams of the mixed valent perovskite manganite oxides. Charge and orbital instabilities, Jahn-Teller (JT) interaction, electron-lattice, AFM super exchange interaction, charge-orbital ordering are also should be considered to explain properly the magnetic nature of mixed valency manganite [7-8].

2.   Structural Properties

In case of ideal perovskite type structure, the atoms are touching one another where B-O distance = a/2 (a = unit cell parameter) and the A-O distance is (a/√2) and the evaluated relationship: $r_A + r_O = \sqrt{2}(r_B + r_O)$, but some of the experimental results show that cubic structure was still have seen in ABO_3 type perovskites and this equation is not correctly followed. After such controversy Goldschmidt [6] represented tolerance factor (t), defined by the following equation:

$$t = \frac{r_A + r_O}{\sqrt{2(r_B + r_0)}}$$

(1)

which is appropriate for room temperature experiential ionic radii. Although for an ideal structure of perovskite, tolerance factor is unity, this structure is also found for different t-values (0.75 < t < 1.0). Generally, at high temperature and tolerance factor near 1, shows an ideal cubic perovskite structure [10].

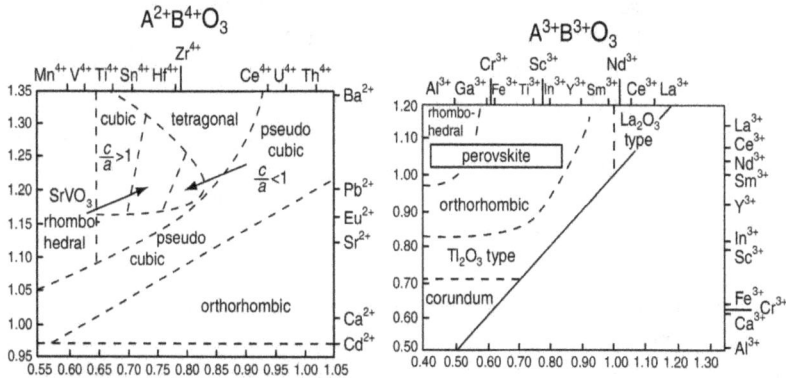

Fig. 4 The effect of ionic size of A- and B-site cations on the observed distortions of the perovskite structure [11].

In perovskite-type compounds, the value of tolerance factor (t) found as more than 1 unity. The tolerance factor value of ABO_3 ideal cubic structure is found to be in the range from 0.89 to 1. Fig. 4 represents the crystal combinations for $A^{2+} B^{4+}O_3$ and $A^{3+}B^{3+}O_3$ groupings, and these are associated with deviation from the ideal structure [11]. In this review chapter, an attempt is made to highlight $A^{3+}B^{3+}O_3$ type perovskite where the A- and B-site is occupied by rare earth and manganese respectively. Based on the obtained results in current research we have tried to compare A-site doping effects on the crystal structure. Recent research with several new instruments discovered that different factor influences on the crystal structure in perovskites oxide materials. Although, size factor and ionic charge are mainly responsible for crystal structure (Fig. 4) but the synthetic

Eng. Magnetic, Dielectric and Microwave Properties of Ceramics and Alloys   Materials Research Forum LLC
Materials Research Foundations **57** (2019) 149-174   doi: https://doi.org/10.21741/9781644900390-7

condition, size, oxygen content, atomic ordering can change the structural arrangement of perovskite oxides.

A-site substituted $R_{1-x}A_xMnO_3$ manganite perovskite oxides were extensively studied due to their interesting behavior depends on the shape, size, oxygen content, synthetic conditions and ordered-disordered phenomenon. The homogeneity range was extended using a "two-steps" synthesis method for the perovskites $R_{1-x}Ba_xMnO_3$ up to x = 0.60 for $Pr_{1-x}Ba_xMnO_3$ and x = 0.50 for $La_{1-x}Ba_xMnO_3$ [12] where lanthanum content samples with the range of $0.20 \leq x \leq 0.40$ reported as rhombohedral phase with *R3hc* space group [13].

Depend on doping concentrations of barium $Pr_{1-x}Ba_xMnO_3$ crystalized at different orthorhombic phases at room temperature where x < 0.20 with *Pbnm*space group shows a double tilting of the MnO_6octahedra but 0.20 < x < 0.40 shows *Ibmm*symmetry with only one tilting of the MnO_6octahedra [14]. A transition to a tetragonal structure with another tilting of $Pr_{1-x}Ba_xMnO_3$ as a rhombohedral structure with the R3c symmetry which was prepared via molten salt synthesis [22]. (0.20 < x < 0.40) reported at low temperature [14]. Although, $Pr_{1-x}Ba_xMnO_3$ ($0.55 \leq x \leq 0.60$) and $La_{1-x}Ba_xMnO_3$ ($0.45 \leq x \leq 0.50$) reported as cubic disordered perovskite but variations of Ba-doping concentrations as $Pr_{1-x}Ba_xMnO_3$ ($0.40 \leq x \leq 0.50$) and $La_{60}Ba_{40}MnO_3$ phase shows a double perovskite type cell [12].

ED patterns of the perovskite $(La, Pr)_{1-x}BaxMnO_3$ shows in Fig. 5(a-e) [12]. Using hydrothermal method Jeffrey et al. prepared [15] single-crystalline nanocubes of $La_{1-x}Ba_xMnO_3$ (x = 0.3, 0.5, and 0.6). These nanocubes crystallized as a pseudo-cubic perovskite structure and the particle sizes formed from 50-100 nm depends on the dopant level [15]. Transmission electron microscopy (TEM) and high-resolution TEM image of $La_{0.5}Ba_{0.5}MnO_3$ and 30-nm $La_{0.7}Ba_{0.3}MnO_3$ nanocube respectively show in Fig. 3(g-f) [15]. Raveau et al. [16] synthesized $La_{1-x}Ba_xMnO_3$ ($0.4 \leq x \leq 0.9$) system in air and found two sets of peaks characteristic of the hexagonal and cubic perovskites. Another research group synthesized 40 to 70 nm diameter content $La_{1-x}Ba_xMnO_3$ nanoparticles using an ionic liquid route and deposited onto mesoporous carbon films, where $LaMnO_3$ was indexed to a cubic (*Pm3m*) phase and the mixed composition samples were indexed to the rhombohedral space group (*R3cH*) in a hexagonal setting [17, 18].

For $Nd_{1-x}BaxMnO_3$ (0 < x < 0.1) showed O'-orthorhombic distortions due to cooperative Jahn–Teller effect, although x = 0.1 does not shows any distortions and the increasing doping concentrations of Ba gradually decrease of the orthorhombic distortion in the crystal structure [16]. The origin of such distortion arises from a mismatch between ionic sizes of different ions which can be expressed as tolerance factor. The reported oxidized and reduced samples of $Nd_{1-x}Ba_xMnO_{3-y}$ ($x > 0.2$, $y < 0.25$) reported as pseudocubic type perovskite structure and the unit cell volume described as increases evidently with increasing of oxygen vacancies which might be due to reducing oxidation state from small Mn^{4+} ions into large Mn^{3+} ions [16]. $Sm_{1-x}BaxMnO_3$ samples reported high distortions in comparison with $Nd_{1-x}Ba_xMnO_{3-y}$. However, the crystal structure distortions of $x = 0.3$ for $Sm_{1-x}BaxMnO_3$ prepared in the air become very small like $Nd_{1-x}Ba_xMnO_{3-y}$ series [16].

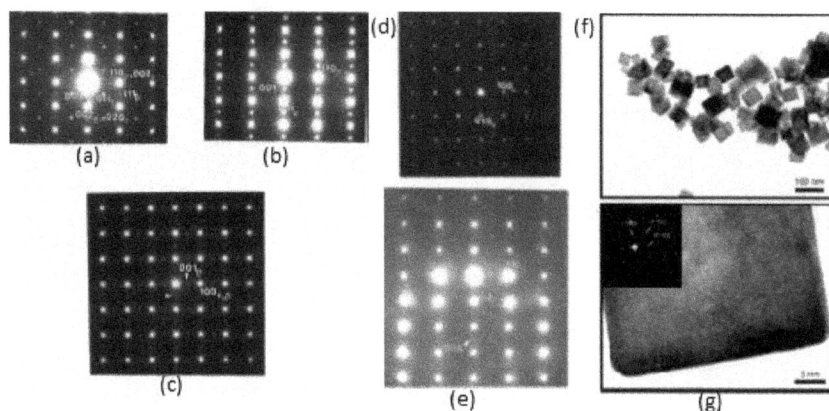

Fig. 5. (A) ED patterns of the perovskite $R_{1-x}BaxMnO_3$. (a) Coexistence of the P-type double perovskite for $Pr_{1-x}BaxMnO_3$ and Imma one for the chemical composition close to x = 0.40. (b) Typical $[110]_P$ ED pattern of $Pr_{0.5}Ba_{0.5}MnO_3$. (c) $[010]_I$ zone of the Imma structure or $[100]_C$ zone of the P-type double-perovskite phase for $Pr_{1-x}BaxMnO_3$. (d) $[001]_P$ of $La_{0.6}Ba_{0.4}MnO_3$. (e) ED patterns for $La_{0.6}Ba_{0.4}MnO_3$ with R3hc and P-type double-cell [6]. (f) TEM image of 20 to 500 nm diameter content $La_{0.5}Ba_{0.5}MnO_3$ nanocubes. Low resolution TEM images for other doping levels are indistinguishable. (g) High-resolution TEM image of a 30-nm $La_{0.7}Ba_{0.3}MnO_3$ nanocube along with a selected area diffraction pattern (inset) [15].

Nakajima et al described [19] the crystal structure of A-site disordered $R_{0.5}Ba_{0.5}MnO_3$ as a primitive cubic perovskite cell and compared with both $RBaMn_2O_6$ and $R_{0.5}A_{0.5}MnO_3$ (A = Sr, Ca). The representation of the most important structural feature of $RBaMn_2O_6$ shown in Fig. 6 a, b where two rock-salt types layers (RO and BaO) make the sandwich with the MnO_2 square sublattice with many different sizes. Therefore, the octahedron MnO_6 is distorted in a non-centrosymmetric manner that both Mn and oxygen atoms are in MnO_2 plane and displaced toward the RO layer (Fig. 6c), in ordered to the rigid MnO_6

octahedron in the A-site disordered perovskite oxide $(R_{1-x}A_x)MnO_3$ (A = Ca and Sr) [19]. According to this explanation, it is easy to say the structural properties of $RBaMn_2O_6$ will be difficult to explain in terms of the basic tolerance factor (t) for $(R_{0.5}A_{0.5})MnO_3$ (A = Ca and Sr) [20]. Fig. 6d and Fig. 6e shows the structure data of Pr contain compounds synthesized by different synthetic conditions and synthetic procedure of ordered-disordered manganite perovskite respectively [19].

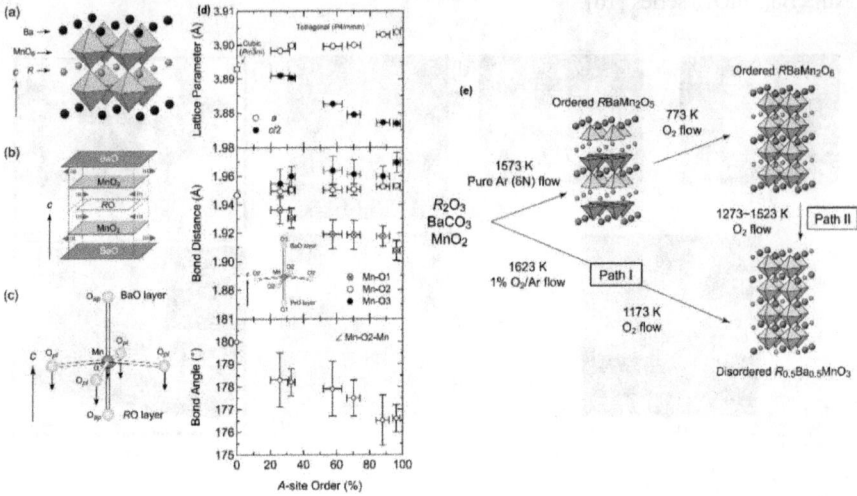

Fig. 6 (a) Crystal structure and (b) structural concept of the A-site ordered manganite
$RBaMn_2O_6$, and (c) an illustration of the distorted MnO_6 octahedron, (d) Structural
parameters at room temperature plotted as a function of the degree of the A-site order
(%) for Pr-compounds, (e) Method of sample preparation [19].

In the case of Sr doped $R_{1-x}Sr_xMnO_3$ samples ($0 \leq x \leq 0.5$), reported by Y. Sakaki et al. [21] which were synthesized using normal ceramics firing procedure and a thermal decomposition method. Although La, Pr, Nd, Sm and Gd contain samples stated as orthorhombic but $La_{1-x}Sr_xMnO_3$ in the range of x = 0.2–0.3 represented as rhombohedral perovskite phase [21]. The degree of orthorhombicity (a /b ratio) has been found to increases from La to Gd and reduces with increasing Sr content [21]. For $La_{1-x}Sr_xMnO_3$ ($0.25 \leq x \leq 0.47$) nanoparticles, the whole study is with different Sr-content samples described as a rhombohedral structure with the R3c symmetry which was prepared via molten salt synthesis [22].

Shlapa et al. reported the effects of synthetic condition on crystal structure and properties [23]. They prepared $La_{1-x}Sr_xMnO_3$ through different methods like as sol-gel precipitation from non-aqueous solution and reversal micro emulsions methods and described that the effects of organic compounds and non-aqueous media and crystallization temperature of synthesized nanoparticles [23]. The morphological studies of $La_{1-x}Sr_xMnO_3$ materials synthesized by Shlapa et al. through the different process are shown in Fig. 7 [23]. The reported $Sm_{1-x}Sr_xMnO_3$ $(0.2 \leq x \leq 0.5)$ [24], nanocrystalline Pr-deficient $Pr_{1-x}Sr_xMnO_{3-\delta}$ [25], $Gd_{1-x}Sr_xMnO_3$ $(0.2 \leq x \leq 0.5)$ [26] $Eu_{1-x}SrxMnO_3$ $(0.2 \leq x \leq 0.5)$ [27] manganites also reported by several research group as orthorhombic phase with *Pbnm* space group.

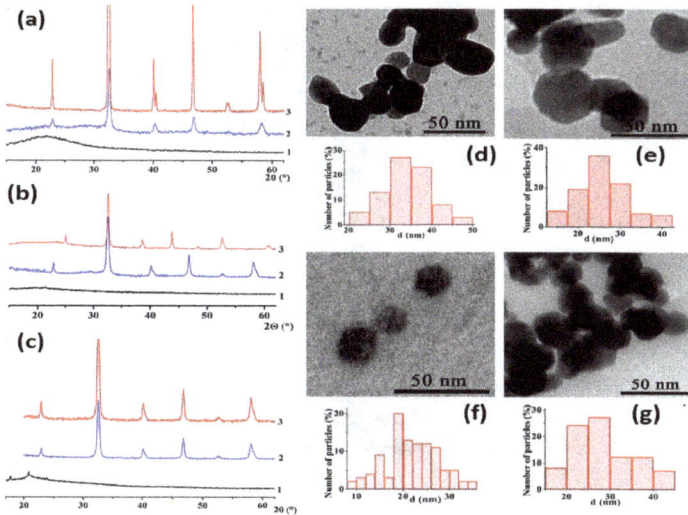

Fig. 7 XRD data for $La_{1-x}Sr_xMnO_3$ nanoparticles, synthesized via sol-gel method (a), by precipitation from Diethylene glycol (DEG) solution (b), and by precipitation from reversal microemulsions (c): 1 denotes 200 °C, 2 denotes 600 °C, and 3 denotes 800 °C, TEM images and particle size distributions of $La_{1-x}Sr_xMnO_3$ nanoparticles synthesized via sol-gel method (d), by precipitation from DEG solution (e), and by precipitation from reversal microemulsions based on Triton X-100 (f) and etyltrimethylammonium bromide (g) [23].

While, calcium content nanocrystalline $La_{1-x}Ca_xMnO_3$ $(0.0 \leq x \leq 0.5)$ reported as cubic symmetry which were prepared by low temperature, self-propagating combustion synthesis procedure [28] but depending on oxygen content it can be crystallized as an orthorhombic phase [29]. Bulk phase of $La_{1-x}Ca_xMnO_3$ $(0.7 \leq x \leq 1)$ with high

Eng. Magnetic, Dielectric and Microwave Properties of Ceramics and Alloys   Materials Research Forum LLC
Materials Research Foundations 57 (2019) 149-174   doi: https://doi.org/10.21741/9781644900390-7

concentration of Ca, also reported as orthorhombic phase [30, 31]. Gonzalez-Calbet et al. [32] studied different types of $La_{0.5}Ca_{0.5}MnO_{3-\delta}$ samples with various δ values using ED and HR-electron microscopy (Fig. 8). Depends on δ values, orthorhombic phase described up to $\delta = 0.25$, and $\delta = 0.5$ reported as the ordering of anionic vacancies with isostructural of Sr_2FeO_5 [32].

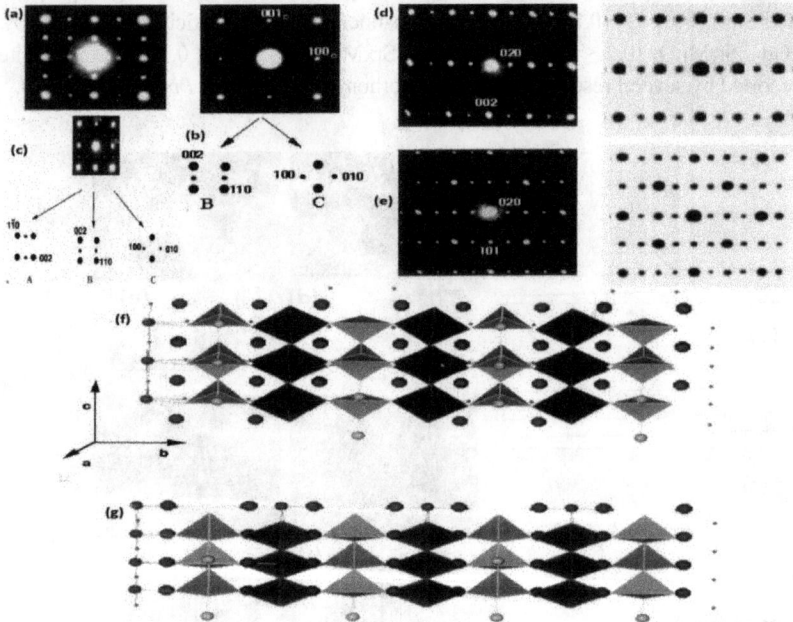

Fig. 8 (a) SAED pattern corresponding to $La_{0.5}Ca_{0.5}MnO_3$ along the $[001]_c$ axis. (b) SAED pattern corresponding to another area of the same. (c) SAED pattern corresponding to $La_{0.5}Ca_{0.5}MnO_{2.75}$ along $[001]_c$ axis. (d, e) Experimental and simulated SAED patterns corresponding to $La_{0.5}Ca_{0.5}MnO_{2.5}$ along [100] and [101] zone axes. (f, g) Structural projections of $La_{0.5}Ca_{0.5}MnO_{2.5}$ along [100] and [101] orientation [32].

Jirak et al. reported the orthorhombic phase of $Pr_{1-x}Ca_xMnO_3$ in the whole range of $0.1 \leq x \leq 0.9$ where orthorhombic distortion noticed in the range of $0.1 \leq x \leq 0.5$ [33]. At low temperature $Pr_{1-x}Ca_xMnO_3$ shows four regions of different symmetry: orthorhombic; $0.1 \leq x \leq 0.3$, pseudotetragonal compressed; $0.3 \leq x \leq 0.75$, pseudotetragonal elongated; $0.75 \leq x \leq 0.9$, and pseudocubic; $0.9 \leq x \leq 1$ [33]. The reported $Pr_{0.5}Ca_{0.5}MnO_3$ in the nano range also shows orthorhombic phase [34]. $Nd_{1-x}Ca_xMnO_3$ ($0 \leq x \leq 0.6$) reported as

Eng. Magnetic, Dielectric and Microwave Properties of Ceramics and Alloys   Materials Research Forum LLC
Materials Research Foundations **57** (2019) 149-174   doi: https://doi.org/10.21741/9781644900390-7

orthorhombic phase, [35, 36] and $Nd_{0.5}Ca_{0.5}MnO_3$ nanocrystalline also shows the orthorhombic structure [37]. In the $La_{1-x}Ca_xMnO_3$ system, it has been reported the coexistence of O'- and O-orthorhombic structures in the range $0.1 \leq x \leq 0.15$ [38]. The conflicting is the case with the elimination of the Jahn-Teller distortions in $Nd_{1-x}Ca_xMnO_3$ where the O'-orthorhombic structure is conserved up to x = 0.5 and has not seen the coexistence of the O'- and O-phases [39]. It has been found from experimental results that the optimum concentration of Ca at which the removal of the O'-orthorhombic distortions arises, at that Ca-concentration O'-orthorhombic distortions increase with reducing the radius of rare earth.

The crystal structure of $Nd_{1-x}Ca_xMnO_3$ manganites in the range of $(0.3 \leq x \leq 0.5)$ has been reported using electron diffraction and neutron diffraction where x = 0.3, 0.4 and 0.5 described as orthorhombic $GdFeO_3$-type structure at room temperature [40]. Authors found the low-temperature polymorph for x = 0.5 in orthorhombic symmetry but monoclinic symmetry for x = 0.4 and 0.3 [40]. Ca doped $Gd_{1-x}Ca_xMnO_3$ and $Eu_{1-x}Ca_xMnO_3$ $(0.1 \leq x \leq 0.5)$ also stated as the orthorhombic phase at room temperature where unit cell volume decreases with increasing doping percentage of calcium [41, 42]. From the above description, one can say that doping of alkali metals in A-site of the perovskite is highly responsible for its crystal structure. Sometimes this doping influences the oxygen content and sometimes to make it ordering. Monitoring oxygen content in perovskite oxides is one of the most crucial steps for tuning their functionality because there are strong relationships between electrons, spins and lattices—stemming from strong hybridization between transition metal d and oxygen p orbitals. Thus, tuning of the crystal structure of perovskite will help to modify the properties of materials for advanced technological applications [43-45].

3.   Magnetical property

Magnetic properties of the perovskite materials can be explained on the basis of either free electron band theory or modified ionic model. Group of metallic phases parallel to the metallic elemental magnets exited in perovskites might be treated as free electron relations. At lower temperatures, strong interaction between the electron's spin lead to a parallel alignment which is responsible for ferromagnetic nature of the perovskite. However, these perovskites convert at higher temperatures into metal/semi-metallic-like Pauli paramagnetic compounds. There is no temperature dependence for the magnetic susceptibility. There is a strong correlation between the electrons in perovskites; however, these elections do not perform as a classical electron gas. For the 3d, 4d and 4f elements, the nature of strong correlation is vital. The collective behaviour of the conduction electrons leads to extensive alteration in half-metals, metal-insulator transitions, CMR and high-temperature superconductivity. Because of the assimilation of paramagnetic positive ions within the perovskite type structure, many perovskites are magnetic. Most significant magnetic types are transition metal cations and lanthanide ions which have partially filled d and f-orbitals. A central feature of superexchange of perovskites is that electron transfer does not take place. Therefore, possible superexchange interactions between cations are present in octahedral and tetrahedral positions. In this section, only those magnetic properties of perovskites are described in some way to be emphasized, or at least unique with suitable examples.

3.1   Phase transitions and magnetoresistance of barium-based manganites

The perovskites $Ln_{1-x}Ba_xMnO_3$ (when x is 0.60) and $Pr_{1-x}Ba_xMnO_3$ (when x is 0.50) were synthesized by two step synthesis mode [19]. The perovskite with $Ln_{1-x}A_xMnO_3$ formula with Ln = Pr and A = alkali metals, extends the homogeneity range of these perovskites. The Curie temperature is obtained 362 K for the sample $Ln_{1-x}Ba_xMnO_3$ [12]. $La_{0.66}Ba_{0.33}MnO_3$ with thin films phase show a large magnetoresistance and resistance ratio R_0/R_H found 50% at 290 K in the presence of 5T external magnetic field [12]. The obtained sample shows the transition from ferromagnetic metallic-to-paramagnetic insulating with rising temperature. The graph shows resistance ratio versus temperature for $Pr_{1-x}Ba_xMnO_3$ phases [12]. Fig. 9 shows the temperature dependence of CMR [12].

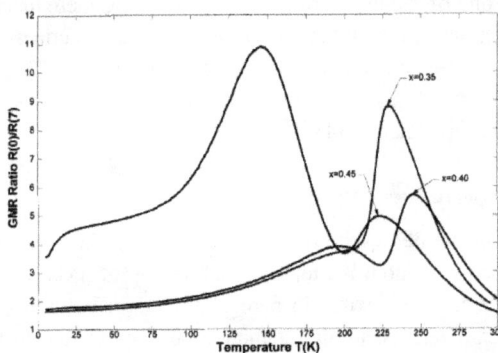

Fig. 9 CMR as a function of Temperature [12].

3.2   Magnetic ordering and granularity effects in $La_{1-x}Ba_xMnO_3$

The magnetic nature of the bulk phase for $La_{1-x}Ba_xMnO_3$ perovskite samples with the composition range (0<x<1), and temperature range ~4.2–775 K, and magnetic field strength ~1G–50 Kg was studied [46]. The moment of $LaMnO_{3+\delta}$ is very sensitive on the processing conditions and the report showed its lowest value of 0.006 μ_B per manganese ion where the magnetic nature of $BaMnO_3$ exhibited antiferromagnetic ordering lower 150 K [46]. The recent research on pseudocubic compounds of the type $La_{2/3}Ba_{1/3}MnO_3$ and $La_{2/3}Ca_{1/3}MnO_3$ observed the huge negative magnetoresistance in hole substituted manganites [47].

3.3   The nature of vacancy-free $La_{1-x}Ba_xMnO_3$

Ba-substituted $LaMnO_3$ with stoichiometric oxygen content was determined using thermogravimetric analysis [48]. Single-phase and vacancy-free barium doped $La_{1-x}Ba_xMnO_3$ have been prepared in the range of x = 0.12– 0.24 using quenching from 1100–1450 °C in the air [48]. Authors claimed the largest magnetoresistive effect for x=0.22 sample at room temperature. Different synthetic conditions of $La_{0.83}Sr_{0.17}MnO_3$ single crystals such as annealed at numerous temperatures and oxygen pressures showed huge variations of magnetic and transport properties. ZFC and FC measurements were reported i.e., temperature vs magnetization of the samples from 5 to 390 K in a dc applied field of 0.002–7 T [48]. The resistivity of field-cooled x =0.22 substituted sample at the different applied field is shown in Fig. 10 [48]. According to the result of the magnetic moments, it was confirmed that orbital angular momentum was quenched for vacancy-free manganite. Temperature dependence of resistance and magnetization vs applied field for Ba-substituted $LaMnO_3$ is given in Fig. 10.

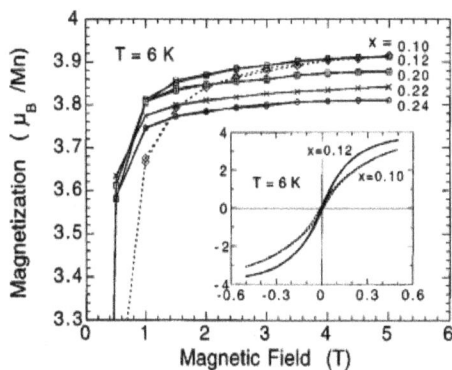

Fig. 10 Resistance as a function of temperature and Magnetization as function of Applied Field. [48].

3.4   Wigner crystal in $R_{1-x}A_x$MnO$_3$ manganite

The expression could be written of double-exchange model of long-range Coulomb force as [$V_{ij} = V_0 a_0 / r_{ij}$] where a_0 is lattice constant. This is used to inspect the electronic phase transition in R$_{1-x}$A$_x$MnO$_3$ samples (Fig. 11) [49]. The interaction strength V_0 is bigger than optimum value, the uniform ferromagnetic state becomes unstable near the composition x=0.5 and the Wigner crystal state was dynamically more satisfactory. The more stable state of spin and charge ordered was found for x=0.5 composition sample than the uniform ferromagnetic state when the force strength becomes bigger than a certain optimum value. The phase diagram found can explanation for the crucial features of experimental observations of manganite perovskites.

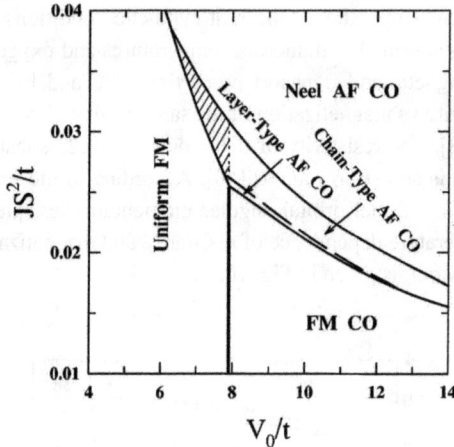

Fig. 11 Electronic phase transition in R$_{1-x}$A$_x$MnO$_3$ compounds [49].

3.5   Transport and magnetic properties of manganite thin films

The thin film fabrication of R$_{1-x}$A$_x$MnO$_{3-\delta}$ (R=La, Nd, A=Ce) perovskite sample using pulsed laser deposition (PLD) method at ambient atmosphere and the study of transport and magnetic properties were reported [50]. Increasing the oxygen pressure leads to an increase in Curie temperature and metal-insulator transition peak appeared near 200 K. The study of detected thermopower that electron-holes are the main carriers in the Ce-doped films i.e. p-type conduction.

Eng. Magnetic, Dielectric and Microwave Properties of Ceramics and Alloys   Materials Research Forum LLC
Materials Research Foundations **57** (2019) 149-174   doi: https://doi.org/10.21741/9781644900390-7

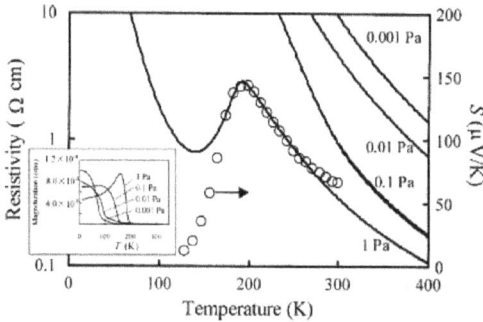

Fig.12 Resistance and Seebeck coefficient vs temperature of $R_{1-x}A_xMnO_{3-\delta}$ (R=La, Nd, A=Ce) thin films [50].

The experimental investigations of manganites perovskite oxide consistently exposed the existence of metal-insulator and ferromagnetic transition and CMR effect [7, 8]. Fig. 12 shows the resistance and Seebeck coefficient vs temperature of $R_{1-x}A_xMnO_{3-\delta}$ (R=La, Nd, A=Ce) thin films [50].

3.6   Magnetic phase diagram of the manganites

The reported single crystal of $Ln_{1-x}BaxMnO_3$ where (Ln=Nd, x=0:23) and (Ln=Nd, Sm 0 < x <0:44) exhibited interesting magnetization and magnetoresistance. Generally, $Ln_{1-x}Ba_xMnO_3$ showed very small crystal structure distortions with a broad range of the lanthanide ionic radii [18]. Ba-doped Nd-content manganite reported the increasing magnetization and magnetic ordering temperature. The doping of ferromagnetic samples such as EuO, $Cd_2Cr_2Se_4$ and $Tl_2Mn_2O_7$ in $Nd_{1-x}Ba_xMnO_3$ system enhanced the magnetoresistance properties and the maximum resistivity showed just below the Curie point. In this system, the magnetoresistance was noticed 30–60% around the Curie point with its negative value.

3.7 $La_{1-x}Ba_xMnO_3$ nanocubes with adjustable doping levels

The doping level is important for the interesting magnetic properties and also depends on size-dependent growth of magnetism, CMR and nanoscale phase. Fig. 13 shows the applied field dependence magnetization at 10 K where nanocube showed clear ferromagnetic switching characteristics [15]. All the $La_{1-x}Ba_xMnO_3$ samples show a minor reduction of saturation magnetization and suppression the temperature of phase transition compared to bulk counterparts.

Eng. Magnetic, Dielectric and Microwave Properties of Ceramics and Alloys   Materials Research Forum LLC
Materials Research Foundations **57** (2019) 149-174   doi: https://doi.org/10.21741/9781644900390-7

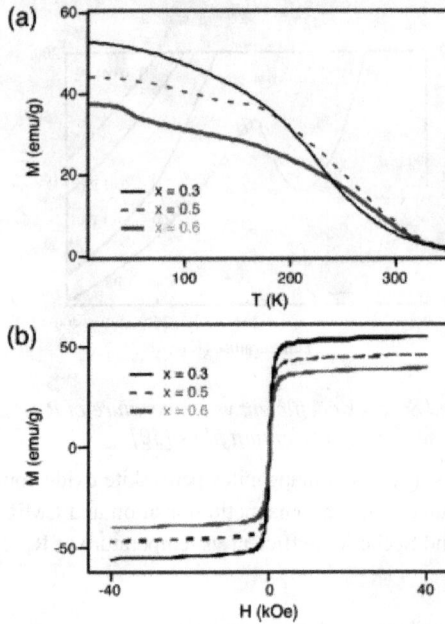

Fig. 13 Magnetization of $La_{1-x}Ba_xMnO_3$ nanocubes at temperature T = 10 K and T = 350 [15]

The study of magnetic properties of $La_{1-x}Ba_xMnO_3$ nanocubes as a function of the applied magnetic field at 10 k have shown in Fig. 13b. The collective results of ZFC, FC and MH clearly demonstrate that all the sample have ferromagnetic to paramagnetic phase transitions with increasing temperature.

3.8 $La_{1-x}Sr_xMnO_3$ nanoparticles and properties

Usually, with reducing the size from bulk to nanoscale are highly responsible for the tuning of structural and magnetic properties. The annealing condition is directly connected with the size of manganite perovskite oxide as well as magnetic parameters especially on the magnetization. There are several methods to obtain weakly agglomerated nanoparticle of manganite. Nowadays, the sol-gel method is one of the promising methods for use as hyperthermia inducers. The magnetic parameters are summarized in Table 1 [23].

Table 1 Sizes and Magnetic Parameters for LSMO nanoparticles syntheised via different methods [23].

Method of Synthesis of $La_{1-x}Sr_xMnO_3$	Synthesis routes			
	Sol-gel	Precipitation from DEG Solution	Micremulsion based on Triton X-100	Micremulsion based on CTAB
Average Particle Size	35±5	25±5	19±5	26±8
Coericive Force H, Oe at 300 K	11.8	3.9	4.2	5.8
Magnetisation Saturation at 300 K	59	49	32	45
Blocking temperature K	~315	~290	~270	~290
SLP (W/g)	38	15	2	21

3.9   Oxygen vacancies ordering in manganite

The clear evidence of orthorhombic perovskite-related phase was recently studied of $La_{0.5}Ca_{0.5}MnO_{3-\delta}$ system using ED and HR-electron microscopy [32]. The paramagnetic moment of several samples with $0 < \delta < 0.5$ illustrations a clear increase as δ rises, as is the case for studied value detected from magnetic susceptibility of $\delta = 0$ to 0.5. The experimental paramagnetic moment was found to be 0.5 kB which was higher than the calculated ones. The mostly ferromagnetic interaction (h= 254 K) was changed to short range antiferromagnetic (h =155 K) with the changes of $\delta = 0$ to 0.5 [32].

3.10   Charge, Spin, and Orbital Ordering in manganite

The mixed valence transition metal content oxide usually shows the charge ordering phenomenon where ordering is done by different charge cations on exact lattice site with unfavorable electrons and cations hopping. This charge-ordering transition allows increasing the electrical resistivity due to the change in crystal symmetry. Fe_3O_4 shows charge ordering where it rises at a temperature below spin ordering [51]. The discovery ofcuprate superconductors, it was given high attention on charge and spin ordering in real space in manganite perovskite oxides [52].

The findings of CMR in perovskite manganates attracted researchers on it to further research [8, 43-44]. The double exchange between the Mn^{3+} $(t_{2g}^3 e_g^1)$ and Mn^{4+} $(t_{2g}^3 e_g^0)$ ions could be the possible explanation of CMR and related properties of manganite system [7]. According to the mechanism, ferromagnetic alignment of partially filled e_g orbitals of adjacent manganese ions is directly correlated to rate of hopping electrons which influence to increase the metal-insulator transition at ferromagnetic Curie temperature. The materials become metallic in the ferromagnetic phase (T < Tc) where insulator in the paramagnetic phase (T > Tc). The Mn^{3+} and Mn^{4+} charge ordering ions participates with double-exchange interaction and endorses insulating behaviour and antiferromagnetic because of Mn^{3+}-O-Mn^{3+} interaction mechanism and Mn^{4+}-O-Mn^{4+} superexchange interaction mechanism through e_g orbitals.

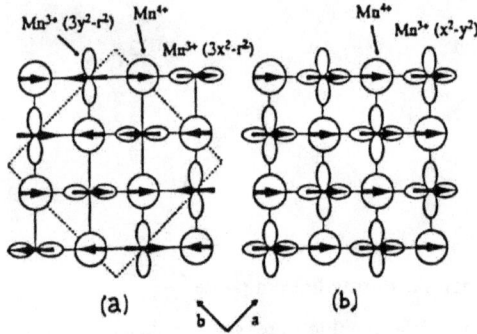

Fig. 14 Charge, spin, and orbital ordering in (a) CE-type and (b) A-type AFM Ln₁₋
$_xA_xMnO_3$. In part a, the broken line shows the unit cell for the CE-type AFM CO order.
Mn^{4+} ions are shown by circles [54].

The general formula of the composition $Ln_{1-x}A_xMnO_3$ is highly interesting because of spin and orbital ordering which allow to the rise of attractive properties [53]. Generally, larger size of lanthanides and divalent A ions such as La and Sr respectively favour ferromagnetism and metallicity, whereas the smaller size of alkali metals combination like (La, Ca, or Pr, Ca) prefer charge ordering. The lattice distortion of d_z^2 orbital also develops long-range ordering which is accompanied with spin ordering. Some manganites show the higher temperature of charge ordering than spin ordering ($T_{CO} > T_N$) and some manganites at the same temperature ($T_{CO} = T_N$). Generally, it is found that orbital ordering arises without T_{CO} in the A-type antiferromagnetic manganates where manganates perovskite system shows CE-type antiferromagnetic (Fig.14) nature along with orbital ordering.

Eng. Magnetic, Dielectric and Microwave Properties of Ceramics and Alloys   Materials Research Forum LLC
Materials Research Foundations **57** (2019) 149-174   doi: https://doi.org/10.21741/9781644900390-7

Conclusion

The review describes the basic fundamentals of the perovskites. The effect of the tolerance factor and ionic size of A- and B-site cations were illustrated with suitable examples of the perovskite. The various techniques used to alter the structure of the perovskites were summarized from the review. The existence of the magnetic property in the perovskite was reviewed on the basis of the free electron band theory. The influence of temperature, synthesis methods and the ordering of the oxygen on CMR and transition temperature were explained from the review.

References

[1]   Feng Jia-Feng, Zhao Kun, Huang Yan-Hong, Zhao Jian-Gao, Han Xiu-Feng, Zhan Wen-Shan and Wong Hong-Kuen, Chinese Physics, 14(9), (2005) 1879-03. https://doi.org/10.1088/1009-1963/14/9/034

[2]   Stefano Curtarolo, Gus LW Hart, Marco Buongiorno Nardelli, Natalio Mingo, Stefano Sanvito, and Ohad Levy, Nature materials, 12(3) (2013) 191. https://doi.org/10.1038/nmat3568

[3]   S. Jin, T. H. Tiefel, M. McCormack, R. A. Fastnacht, R. Ramesh and L. H. Chen, Science, 264 (1994) 413-415. https://doi.org/10.1126/science.264.5157.413

[4]   Graham M. McNally, Ángel M. Arévalo-López, Padraig Kearins, Fabio Orlandi, Pascal Manuel, and J. Paul Attfield, Chem. Mater., 29 (20) (2017) 8870–8874. https://doi.org/10.1021/acs.chemmater.7b03556

[5]   Aslam Hossain, Prasanta Bandyopadhyay and Sanjay Roy, Journal of Alloys and Compounds, 740 (2018) 414-427. https://doi.org/10.1016/j.jallcom.2017.12.282

[6]   Megaw, H. D. Proc. Phys. Soc. 58 (1946)133. https://doi.org/10.1088/0959-5309/58/2/301

[7]   Zener, C. Phys. ReV., 82 (1951) 403. https://doi.org/10.1103/PhysRev.82.403

[8]   R. von Helmolt, J. Wecker, B. Holzapfel, L. Schultz, and K. Samwer, Phys. Rev. Lett. 71 (1993) 2331–2333. https://doi.org/10.1103/PhysRevLett.71.2331

[9]   JA Alonso, M. J. Martinez-Lope, M. T. Casais, and M. T. Fernandez-Diaz, Inorganic chemistry 39(5) (2000) 917-923. https://doi.org/10.1021/ic990921e

[10]   Khattak, C. P., and Wang, F. F. Y. In Handbook of the Physics and Chemistry of Rare Earths; Gschneider, K. A. Jr.; Eyring, L., Eds.; North-Holland Publisher: Amsterdam, (1979) 525

[11]   T. Ishihara (ed.), Perovskite Oxide for Solid Oxide Fuel Cells, Fuel Cells and Hydrogen Energy

[12]   A. Barnabe, F. Millange, A. Maignan, M. Hervieu, and B. Raveau, Chem. Mater.,10 (1998) 252-259. https://doi.org/10.1021/cm9704084

[13]   Radaelli, P. G. Marezio, M. Hwang, H. Y., and Cheong, S. W. J. Solid State Chem., 122 (1996) 444-447. https://doi.org/10.1006/jssc.1996.0140

[14]   Jirak, Z. Pollert, E. Andersen, A. F. Grenier, J. C. and Hagenmuller, P. Eur. J. Solid. State Inorg. Chem., 27 (1990) 421-433

[15]   Jeffrey J. Urban, Lian Ouyang, Moon-Ho Jo, Dina S. Wang, and Hongkun Park, Nano Lett., 4 (8) 2004. https://doi.org/10.1021/nl049266k

[16]   B Raveau, C Martin, A Maignan and M Hervieu, J. Phys. Condens. Matter 14 (2002) 1297–1306. https://doi.org/10.1088/0953-8984/14/6/316

[17]   G. P. A. Gobaille-Shaw, V. Celorrio, L. Calvillo, L.J. Morris, and G. Granozzi D. J. Fermín, Chem Electro Chem, 5 (2018)1–7. https://doi.org/10.1002/celc.201800052

[18]   I O Troyanchuk, D D Khalyavin, S V Trukhanov and H Szymczak, J. Phys. Condens. Matter 11 (1999) 8707–8717. https://doi.org/10.1088/0953-8984/11/44/309

[19]   Tomohiko Nakajima, Hideki Yoshizawa and Yutaka Ueda, Journal of the Physical Society of Japan 73(8) (2004) 2283–2291. https://doi.org/10.1143/JPSJ.73.2283

[20]   C. N. R. Rao and B. Raveau: Colossal Magnetoresistance, Charge Ordering and Related Properties of Manganese Oxides (World Scientific, Singapore) (1998)

[21]   Y. Sakaki, Y. Takeda, A. Kato, N. Imanishi, O. Yamamoto, M. Hattori, M. Iio, and Y. Esaki, Solid State Ionics 118 (1999) 187–194. https://doi.org/10.1016/S0167-2738(98)00440-8

[22]   M. Kačenka, O. Kaman, Z. Jirák, M. Maryško, P. Veverka, M. Veverka, and S. Vratislav, Journal of Solid State Chemistry 221 (2015) 364–372. https://doi.org/10.1016/j.jssc.2014.10.024

[23]   Yulia Shlapa, Sergii Solopan, Anatolii Belous and Alexandr Tovstolytkin, Nanoscale Research Letters 13 (2018)13. https://doi.org/10.1186/s11671-017-2431-z

[24]   Nagaraja B.S., Ashok Rao, Poornesh P, Tarachand, and G. S. Okram, Physica B: Condensed Matter, 523 (2017) 67-77. https://doi.org/10.1016/j.physb.2017.08.027

[25]   B. Arun, M. Athira, V. R. Akshay, B. Sudakshina, Geeta R. Mutta, and M. Vasundhara, Journal of Magnetism and Magnetic Materials, 448 (2018) 322-331. https://doi.org/10.1016/j.jmmm.2017.06.105

[26]   Nagaraja B. S, Ashok Rao, P. D Babu, and G. S. Okram, Physica B Condensed Matter, 479 (2015)10-20. https://doi.org/10.1016/j.physb.2015.09.025

[27]   Nagaraja B. S, Ashok Rao, P. D Babu, and G. S. Okram, Journal of Alloys and Compounds, 683 (2016) 308-317. https://doi.org/10.1016/j.jallcom.2016.05.098

[28]   B. M. Nagabhushana, R. P. Sreekanth Chakradhar, K. P. Ramesh, C. Shivakumara, and G. T. Chandrappa, Materials Chemistry and Physics 102 (2007) 47–52. https://doi.org/10.1016/j.matchemphys.2006.11.002

[29]   S. V. Trukhanov, N. V. Kaspera, I. O. Troyanchuk, M. Tovar, H. Szymczak, and K. Bärner, Journal of Solid State Chemistry, 169 (2002) 85-95. https://doi.org/10.1016/S0022-4596(02)00028-2

[30]   M. Arunachalam, P. Thamilmaran, S. Sankarrajan, and K. Sakthipandi, Physica B: Condensed Matter, 456 (2015) 118-124. https://doi.org/10.1016/j.physb.2014.08.033

[31]   Hideki Taguchi, Journal of Solid State Chemistry, 124 (1996) 360-365. https://doi.org/10.1006/jssc.1996.0250

[32]   J. M. Gonzalez-Calbet, E. Herrero, N. Rangavittal, J. M. Alonso, A J. L. Martinez and M. Vallet-Regi, Journal of Solid State Chemistry 148 (1999) 158-168. https://doi.org/10.1006/jssc.1999.8441

[33]   Z. Jirak, S. Krupicka, and Z. Simsa, Journal of Magnetism and Magnetic Materials 53 (1985) 153-166. https://doi.org/10.1016/0304-8853(85)90144-1

[34]   S. S. Rao and S V Bhat, J. Phys.: Condens. Matter 22 (9) (2010) 116004. https://doi.org/10.1088/0953-8984/22/11/116004

[35]   K. Raju, K.V. Sivakumar, and P. Venugopal Reddy, Journal of Physics and Chemistry of Solids 73 (2012) 430–438

[36]   Md. Motin Seikh, Chandrabhas Narayana, A.K. Cheetham, and C.N.R. Rao, Solid State Sciences 7 (2005) 1486–1491

[37]   L. Liu, S. L. Yuan, Z. M. Tian, X. Liu, J. H. He, P. Li, C. H. Wang, X. F. Zheng and S Y Yin, J. Phys. D: Appl. Phys. 42 (4) (2009) 045003. https://doi.org/10.1088/0022-3727/42/4/045003

[38]   I. O. Troyanchuk and S. N. Pastushonok, Phys. Stat. Solidi (a), 115 (1989) 225. https://doi.org/10.1002/pssa.2211150254

[39]   I. O. Troyanchuk, S. N. Pastushonok, O. A. Novitskii and V.I. Pavlov, Journal of Magnetism and Magnetic Materials 124 (1993) 55-61. https://doi.org/10.1016/0304-8853(93)90069-E

[40]   O. Richard, W. Schuddinck, G. Van Tendeloo, F. Millange, M. Hervieu, V. Caignaert and B. Raveau, Acta Cryst., 55 (1999) 704-718. https://doi.org/10.1107/S0108767398012215

[41]   S. Kh. Estemirova, V. F. Balakirev, A. M. Yankin, V. Ya. Mitrofanov, S. A. Uporov, V. M. Kozin, and T. I. Filinkova, Glass Physics and Chemistry, 41 (2) (2015) 224–231. https://doi.org/10.1134/S1087659615020066

[42]   I. O. Troyanchuk and N. V. Samsonenko, Phys. Solid State, 39 (1997) 101. https://doi.org/10.1134/1.1130154

[43]   Aslam Hossain, Sanjay Roy, and K. Sakthipandi, Ceramics International (2018)

[44]   Chahara, K. Ohno, T. Kasai, and M. Kozono, Y. Appl. Phys. Lett. 63 (1993) 1990. https://doi.org/10.1063/1.110624

[45]   S. Sankarrajan, K. Sakthipandi, and V. Rajendran, Materials Research, 15 (4) (2012) 517-521. https://doi.org/10.1590/S1516-14392012005000067

[46]   R.C.Budhani, Chaitali Roy, Laura H. Lewis, Qiang Li, and A. R. Moodenbaugh, Magnetic ordering and granularity effects in La1−xBaxMnO3, Journal of Applied Physics 87 (2000) 2490. https://doi.org/10.1063/1.372208

[47]   G. H. Jonker and J. H. van Santen, Physica Amsterdam, 16 (1950) 337. https://doi.org/10.1016/0031-8914(50)90033-4

[48]   B. Dabrowski, K. Rogacki, X. Xiong, P. W. Klamut, R. Dybzinski, J. Shaffer, and J. D. Jorgensen, Phys. Rev. B, 58 (1998) 2716. https://doi.org/10.1103/PhysRevB.58.2716

[49]   L. Sheng and C. S. Ting, Phys. Rev. B, 57 (1998) 5265. https://doi.org/10.1103/PhysRevB.57.5265

[50]   T. Yanagida, T. Kanki, B. Vilquin, H. Tanaka, and T. Kawai, Thin Solid Films, (486) 122-124. https://doi.org/10.1016/j.tsf.2004.11.210

[51]   Honig, J. M. Proc. Indian Acad. Sci., Chem. Sci., 96 (1986) 391. https://doi.org/10.1007/BF02936294

[52]   Tranquada, J. M. Sternleib, B. J. Axe, J. D. Nakamura, Y. and Uchida, S., Nature, 375 (1995) 561. https://doi.org/10.1038/375561a0

[53]   Rao, C. N. R.; Arulraj, A.; Santosh, P. N.; Cheetham, and A. K. Chem. Mater., 10 (1998) 2714. https://doi.org/10.1021/cm980318e

[54]   C. N. R. Rao, J. Phys. Chem. B, 104 (2000) 5877-5889. https://doi.org/10.1029/1998JE900037

Eng. Magnetic, Dielectric and Microwave Properties of Ceramics and Alloys   Materials Research Forum LLC
Materials Research Foundations **57** (2019) 175-190   doi: https://doi.org/10.21741/9781644900390-8

Chapter 8

Microwave Absorption in Ceramics: Different Mechanisms and its Optimization

Charanjeet Singh[1,a,*], Sukhleen Bindra Narang[2,b], Rajshree Jotania[3,c]

[1] Department of Electronics and Communication Engineering, Lovely Professional University, Phagwara, Punjab, India

[2]Department of Electronics Technology, Guru Nanak Dev University, Amritsar, Punjab, India

[3]Department of Physics, Electronics and space science, School of sciences, Gujarat University, Ahmedabad 380 009, India

[a]rcharanjeet@gmail.com; [a]charanjeet2003@rediffmail.com; [b]sukhleen2@yahoo.com; [c]rbjotania@gmail.com

Abstract

In the current scenario of technological devices, industrial, domestic and military applications involve high speed electronic devices operating in GHz region. The microwave absorbers are the key materials incorporated to mitigate electromagnetic interference emanated from high speed electronic devices. In the present chapter, we have discussed microwave absorption mechanisms which are not much explored analytically as well as comprehensively. The necessary mathematical models responsible for absorption have been elaborated with pertinent mechanisms. Both material science and engineering aspects have been covered to elaborate on the conceptual understanding of optimization of microwave absorptions in any ceramics.

Keywords

Ceramics, Microwave Absorption, Mechanisms, Magnetic Hysteresis Loop

Contents

Eng. Magnetic, Dielectric and Microwave Properties of Ceramics and Alloys   Materials Research Forum LLC
Materials Research Foundations **57** (2019) 175-190   doi: https://doi.org/10.21741/9781644900390-8

1.   Introduction

The exponential rise in information technology is associated with high speed electronic devices working in the microwave and millimeter wave regime. As a result of this, the devices start emanating unwanted electromagnetic radiation also termed as electromagnetic interference (EMI), which interfere with the functioning of nearby conductors/electronic devices through electromagnetic induction. The stray EMI of high speed devices can cause bit error in the digital modulated data received by wireless receivers. The very large scale integration (VLSI) has rendered easy integration of numerous components on single sided and double sided printed circuits boards, Nowadays, printed circuit boards of laptops, synthesizers or supercomputers constitute thin width of numerous copper conductors into order to integrate a large number of electronic components. However, these conductors start radiating (EMI) as an antenna, when the width of copper conductors tracks becomes an integral multiple of a quarter wavelength of high frequency signal passing through the conductors. The microwave absorbers are utilized for attenuating the EMI or stray electromagnetic reflection from military aircraft, tank, radar, etc.

Researchers have investigated magnetic and dielectric ceramics for possible application as a microwave absorber [1-16]. The absorption has been optimized by increasing dielectric and/or magnetic losses through doping of divalent, trivalent, tetravalent ions, etc. Similarly, composites are also being explored by adding different matrix into ceramics such as paraffin wax, polystyrene, graphene, rubber, polyurethane, graphite, etc.

The main objective of absorption enhancement is focused on new structures and losses. However, the relation between static i.e. hysteresis properties and microwave absorption is analyzed qualitatively and a few reports are available for a contribution of loss tangent in impedance matching. Most of the reports have included the discussion of input impedance ($|Z_{in}|$) of the absorber for the purpose of matching with free space impedance ($Z_0 = 377 \, \Omega$). This discussion is not appropriate since Z_0 is a real number and Z_{in} is a complex number having both real as well as complex values.

In view these mentioned rebuttals, the present chapter highlights different mechanisms of absorption and its' optimization. The relation between static properties (DC) and dynamic properties is also enunciated.

2. Microwave absorption

Complex permittivity and complex permeability are prerequisite to look into for understanding the absorption in the ceramics. Both quantities can be written as:

Complex permittivity $\varepsilon = \varepsilon' - j\varepsilon''$

Complex permeability $\mu = \mu' - j\mu''$

Where μ', μ'', ε' and ε'' are permeability, magnetic loss, dielectric constant and dielectric loss respectively.

μ' represents energy storage in the form of the magnetic field while ε' corresponds with energy storage in the electric field, μ'' and ε'' ascribe to energy loss or dissipation. The dipolar polarization is primarily dominant at microwave frequencies in the ceramics.

2.1 Dielectric loss

The dielectric properties are dependent on various types of polarization viz-a-viz interfacial polarization, electronic polarization, atomic polarization, ionic polarization and dipole polarization. At higher frequencies, the interfacial polarization and intrinsic electric dipole polarization mainly contribute to the enhancement of complex permittivity [17]. The difference in the valency between the dopants and main elements converts the valency of parent elements. This gives impetus to the interfacial and intrinsic dipole polarization thereby increasing the complex permittivity.

2.1.1 Dielectric relaxation

It is the delay or lag in the dielectric constant of a material after the input signal is applied. More specifically, it is due to the delay in polarization in the material with the changing electric field of the microwave signal. The dielectric relaxation in the dynamic electric field is analogous to the hysteresis loss in varying magnetic fields.

It is also described as permittivity vs. frequency and various researchers have described it in the form of the Debye equation. Mathematically, it can be represented as:

$$\varepsilon\left(\omega\right) = \varepsilon_\infty + \frac{\Delta\varepsilon}{1+i\omega\tau} \tag{1}$$

Where ε_∞ is the permittivity at the high frequency and $\Delta\varepsilon = \varepsilon_s - \varepsilon_\infty$ where is ε_s the permittivity at the low frequency, τ is the characteristic relaxation time of the material.

2.1.2 Cole-Cole equation

The Cole-Cole equation represents the relaxation model in polycrystalline ceramics or polymers and can be expressed as:

$$\varepsilon\left(\omega\right) = \varepsilon_\infty + \frac{\Delta\varepsilon}{1+i\omega\tau^{1-\alpha}} \tag{2}$$

The exponent α can take along values between 0 and 1. When $\alpha=0$, this model traces back to the Debye model. However, with $\alpha>0$, the dielectric relaxation extends over the large frequency range on a logarithmic ω scale in comparison to the Debye relaxation.

2.1.3 Cole–Davidson equation

$$\varepsilon\left(\omega\right) = \varepsilon_\infty + \frac{\Delta\varepsilon}{(1+i\omega\tau)^\beta} \tag{3}$$

2.1.4 Havriliak–Negami relaxation

Havriliak–Negami relaxation model considers asymmetry and broadness of the dielectric dispersion plots, and can be written as:

$$\varepsilon\left(\omega\right) = \varepsilon_\infty + \frac{\Delta\varepsilon}{(1+i\omega\tau^\alpha)^\beta} \tag{4}$$

Where ε_∞ is the permittivity at the high frequency and $\Delta\varepsilon = \varepsilon_s - \varepsilon_\infty$ where is ε_s the permittivity at the low frequency, τ is the characteristic relaxation time of the material. The exponents are indicative of the asymmetry and broadness of the corresponding spectra. For $\beta=1$, the mentioned equation reduces to the Cole-Cole equation, while $\alpha=1$ translates it to the Cole–Davidson equation.

2.2   Magnetic losses

The magnetic loss constitutes domain wall resonance, eddy current and hysteresis attenuation [17]. It is known that the resonance in the domain wall occurs around 1 GHz. The hysteresis attenuation must be small in the absorber failing it will not be able to follow the signal. As a result of this, the signal will pass through the absorber without attenuation.

2.2.1   Eddy current loss

The eddy currents are related to the skin effect in the material and these currents are more prominent at microwave frequencies. The high frequency carries with rapid variation in microwave signal along with time domain: the fast changing electric field produces a magnetic field of the same variation, the variation in the induced magnetic field further produces electric field or eddy current which opposes the original electric field (current). Thus the induced eddy currents render the hindrance to the source (microwave signal) and considerable attenuation or microwave absorption of the microwave signal is possible.

The existence of eddy current can be evaluated mathematically as:

$$(\mu'')(\mu')^{-2}f^{-1} = 2\Pi\,\mu_0\sigma d^2 \qquad\qquad (5)$$

Let $C_o = (\mu'')(\mu')^{-2}f^{-1}$

Where μ' is the permeability, μ'' is the magnetic loss, μ_0, σ, and d are permeability of vacuum, conductivity and thickness of composition respectively.

The graph is to be plotted for C_o as a function of the investigated microwave frequency region. For the eddy current to exist, C_o should be independent of frequency [17].

2.3   Quarter wavelength mechanism

According to this mechanism when the thickness of ferrite material is equal to the odd integer multiple of wavelength $n\lambda/4$ (where n = 1, 3, 5....) of the microwave signal, it will be absorbed while passing through the material [19].

Mathematically,

$$t = \frac{n\lambda}{4} = \frac{n.c}{4f\sqrt{|\mu_r.\varepsilon_r|}} \tag{6}$$

Where t is the matching thickness, n is an integer, c is the velocity of light, fm is the matching frequency are the matching thickness, an integer, the velocity of light, the matching frequency respectively. and $\mu_r = \mu' - j\mu''$, $\varepsilon_r = \varepsilon' - j\varepsilon''$ have their usual meaning discussed before. For perfect dielectric ceramics, μ' is taken 1 and $\mu''=0$ and similarly for perfect dielectric materials.

According to equation (6), the thickness of absorber can be reduced by frequency and material parameters, while the frequency region for maximum absorption can be tuned by varying the thickness. The absorber thickness is dependent on the product of two parameters complex permittivity and complex permeability, and not their individual values.

To evaluate microwave absorption, transmission line theory elucidates the reflection loss (RL) as a function of the input impedance (Z_{in}). It can be expressed by the relation:

$$RL = 20log \left| \frac{(Z_{in}-Z_o)}{(Z_{in}+Z_o)} \right| \tag{7}$$

Where $Z_o = 377 \ \Omega$ is the characteristic impedance of free space and Z_{in} is the input impedance of a metal backed absorber which can be written as [20]:

$$Z_{in} = Z_o \sqrt{\frac{\mu_r}{\varepsilon_r}} tanh \left[j(\frac{2\pi ft}{c})\sqrt{(\mu_r.\varepsilon_r)} \right] \tag{8}$$

The impedance matching condition is given by $Z_{in} = Z_o$ or ratio of Z_{in}/Z_o must be equal to 1 in order to have ideal absorbing properties.

It can be deduced from equation (7) of RL (microwave absorption) is dependent on Z_{in} (input impedance) of the absorber in equation (8). Thus RL is dependent on 6 parameters viz a viz μ', μ", ε', ε", t and f i.e. material parameters, geometry (thickness) of the absorber and frequency. The material parameters change with frequency, however, are independent of the thickness of absorber. Further, equation (4) also involves the ratio as well as multiplication of complex permeability and complex permittivity, and not the individual values of these two parameters. Therefore, manipulation of the parameters can be done to obtain the same input impedance.

In Fig. 1, plots depict the RL (microwave absorption) versus frequency for simulated thickness in Co-Ga doped M-type Ba-Sr hexagonal ferrite [21]. RL dips in compositions

$x = 0.2$ moves along with the low frequency region with the increment in the thickness and vice-versa. It is attributed to the quarter wavelength mechanism (equation 2) in which frequency and thickness are of reciprocative nature. RL dip of value -29.74 dB has been ascribed in $x=0.2$ at frequency and thickness of 8.28 GHz and 2.0 mm respectively.

For confirming the correlation between quarter wavelength mechanism and observed RL dip, calculated thickness [t_{cal}, equation (6)] is compared with simulated thickness [t_{sim}, equation (8)]. Figure 1 displays t_{cal} or $n\lambda/4$ (with $n=1$) plots as a function of frequency, that value of n is taken which gives t_{cal} is near to t_{sim}. Hence for RL dip $\geq$ -10 or -20 dB, $x=0.2$ owes the existence of quarter wavelength mechanism with t_{cal} close to t_{sim}. Also, this composition has closeness between measured, simulated and calculated thickness of 2.1 mm, 2 mm and 2.06 mm respectively.

2.4   Loss tangent contribution in impedance matching

The impedance matching can be achieved in terms of the synergistic effect associated with dielectric loss tangent (tan δ_ε) and magnetic loss tangent (tan δ_μ). The small value of differences between magnetic loss and dielectric loss tangent encourages the impedance matching, thereby the microwave absorption increases [22]. Fig. 2 demonstrates the relatively better matching in $x = 0.0$, 0.2 and 0.4 than $x = 0.6$, 0.8 and 1.0 and former compositions can have the possibility of enhancement of microwave absorption.

2.4.1 Impedance matching between absorber and medium of propagation

The researchers have focused more on increasing the dielectric and magnetic losses for the enhancement of microwave absorption. However, the first condition for absorption is that signal propagating from in air or vacuum medium should enter the absorber with maximum strength i.e. without reflection: Fig. 3 displays the realization of this case. It is possible if the impedance of absorber is matched with the characteristic impedance of the medium. If there is a impedance mismatch between the two, then part of the signal will be reflected from the absorber, remaining signal will enter the absorber. This problem is illustrated in fig. 3 which displays the three components in signal viz a viz incident, reflected and transmitted. Thus even if absorber is designed with maximum tan δ_ε and tan δ_μ, a portion

Eng. Magnetic, Dielectric and Microwave Properties of Ceramics and Alloys   Materials Research Forum LLC
Materials Research Foundations **57** (2019) 175-190   doi: https://doi.org/10.21741/9781644900390-8

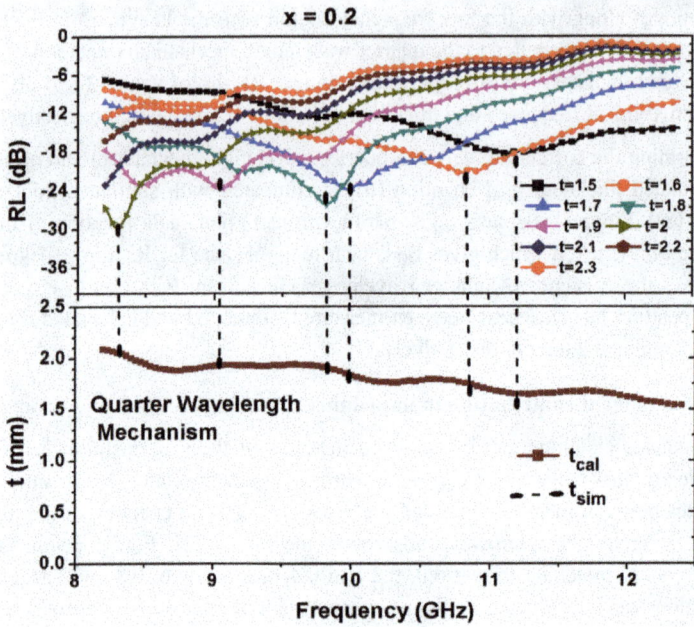

Fig. 1 Dependence of RL, t_{sim} and t_{cal} on frequency in composition $x=0.2$ for various thickness.

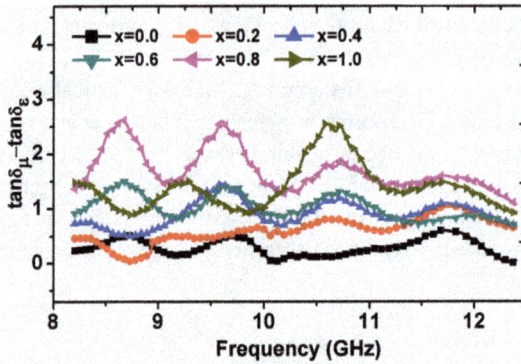

Fig.2 Difference of dielectric loss and magnetic loss tangent (Rights and Permission Springer-Nature) [21].

Eng. Magnetic, Dielectric and Microwave Properties of Ceramics and Alloys   Materials Research Forum LLC
Materials Research Foundations **57** (2019) 175-190   doi: https://doi.org/10.21741/9781644900390-8

of the reflected signal is lost due to reflection and remaining signal is absorbed due to large losses. Therefore, there is a need for an analytical explanation of absorber impedance and characteristic impedance.

Fig. 4 represents the dependence of the highest RL (composition x=0.2) and $|Z_{in}|$ on the frequency at 2 mm thickness. RL in equation (7) will be large with Z_{in} near to Z_0. The horizontal line in the graph represents the characteristic impedance (Z_o = 377 Ω) of air. For impedance matching, Z_{in} is should be equal to Z_o, however, Z_{in} involves complex transcendental equation (8).

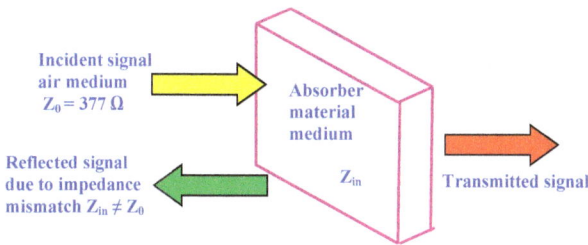

Fig. 3 Impedance mismatch between absorber medium and air medium.

Z_{in} is not a real number like Z_o : $Z_{in} = Z_{real} + j\ Z_{img}$ (Z_{real} is a real part of the impedance and Z_{img} is imaginary part of impedance). The large RL will result from equation (7) when $|Z_{in}|$ is near to Z_o=377Ω. Thus both Z_{real} should be near to 377 Ω and zero respectively and RL decreases if either or both of Z_{real} and Z_{img} higher or lower than 377 Ω and zero respectively.

The plots of Z_{in}, Z_{real} and Z_{img} vs. frequency are shown in figure 4. Among all compositions, there is an observation of maximum RL (-29.74 dB) in composition x = 0.2 with Z_{real} =353.38 Ω and Z_{img} =3.69 Ω. Both the impedances are relatively more close to 377 Ω and zero respectively. On the other side, rest of the compositions owe low RL with Z_{real} or Z_{img} more offset from 377 Ω and/or zero respectively. Furthermore, these compositions have large magnetic losses too. Consequently, composition x=0.2 has a significant contribution of both impedance matching and quarter wavelength mechanism for the microwave absorption.

2.5   Hysteresis parameters

The hysteresis parameters play an important role in the dependence of microwave absorption. The pre-requisite for a good absorber is to have magnetically soft nature i.e.

easily magnetized and demagnetized. As shown in fig. 5, hysteresis loss should be as small as possible: for a large loss, the material will not able get magnetize during the positive cycle and demagnetize during the negative half cycle of the microwave signal. In other words, the material will not be able to follow the microwave signal and the signal will pass as such without absorption.

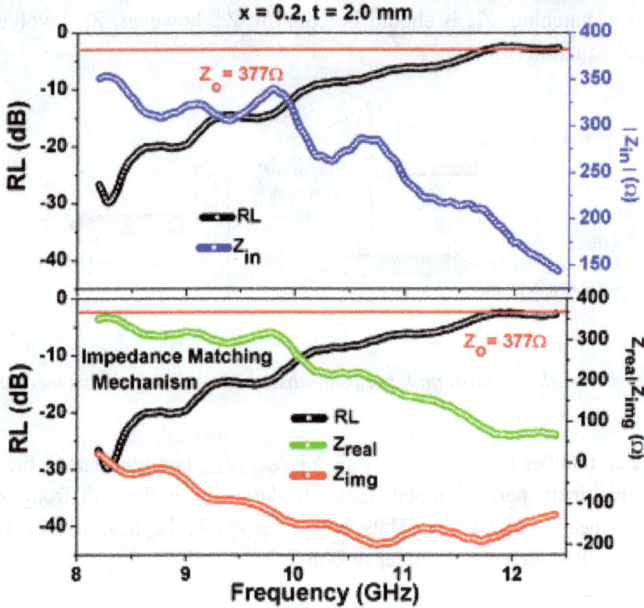

Fig. 4 Dependence of RL dip, Z_{in}, Z_{real} and Z_{img} on frequency in composition x=0.2 at a thickness of 2.0 mm (Rights and Permission Springer-Nature) [21].

Fig. 5 shows hysteresis loops of $Ba_{0.5}Sr_{0.5}Co_xW_xFe_{12-2x}O_{19}$ hexagonal ferrite [23]. The doping of Co^{2+} and W^{4+} causes a reduction in the area of hysteresis loops i.e. hysteresis loss decreases. The fall in coercivity (H_c) is the deterministic factor for narrow hysteresis loop. Furthermore, coercivity is dependent on grain size and anisotropy field (H_a).

The magnetic loss (μ'') is dependent on hysteresis parameters and can be expressed with the following relation:

$$\mu'' = \frac{M_s}{2H_a\propto} \tag{9}$$

Eng. Magnetic, Dielectric and Microwave Properties of Ceramics and Alloys   Materials Research Forum LLC
Materials Research Foundations **57** (2019) 175-190   doi: https://doi.org/10.21741/9781644900390-8

where M_s, H_a and α are saturation magnetization, anisotropy field and coefficient of extinction respectively.

In other words, μ'' [21] varies in accordance with M_s/H_a and input impedance (Z_{in}), which involves μ'' in equation 8, can be tuned by M_s/H_a. Thus there exists a relationship between AC (frequency dependent; complex permittivity and permeability) and DC properties (frequency independent, Hysteresis). As shown in table 1, variation in μ'' [21] followed the variation in M_s/H_a concluding that Z_{in}, and therefore microwave absorption, is dependent on the hysteresis parameters. Therefore mechanisms of M_s and H_a need to be known and primary depend on preferential site occupancy of dopant ions. For example, the increase in M_s is observed with occupancy of doped ions (less magnetic than Fe^{3+}) on spin down sites, whereas it decreases with occupancy on spin up sites. Similarly, H_a carries with a ferromagnetic resonance which is discussed in next section.

2.6   Ferromagnetic resonance

The ferrites possess spontaneous magnetization i.e. they have their own fields in the absence of an external field. With the application of external static magnetic field H, the ferrite will start spinning and will be saturated after the application of the magnetic field. The ferrite electrons will come to rest with their spin axis and magnetic moments parallel to the applied field. Now, when the dynamic magnetic field is applied perpendicular to the static field H, the electrons will start precessing in larger paths until they arrive at the equilibrium precession orbit due to applied magnetic fields and the internal damping.

When the frequency of the dynamic magnetic field is equal to the precession frequency of the electrons in the ferrite, there will be energy transfer from the alternating magnetic field to the precessing electrons in the ferrite. This energy is dissipated in the form of heat thereafter and this phenomenon is called resonance absorption or ferromagnetic resonance (FMR). Therefore, whenever the microwave signal will pass through the ferrite material, ferromagnetic resonance will occur and it is 42 GHz for M-type Ba hexagonal ferrites.

The ferromagnetic resonance can be calculated from the following expression [24]:

$$f_r = \frac{\gamma}{2\pi} H_a \qquad (10)$$

where f_r, γ and H_a are the ferromagnetic resonance frequency, the gyromagnetic ratio and the anisotropy field respectively. Table 1 lists H_a of different compositions and f_r lies outside the investigated frequency region.

185

Eng. Magnetic, Dielectric and Microwave Properties of Ceramics and Alloys   Materials Research Forum LLC
Materials Research Foundations **57** (2019) 175-190   doi: https://doi.org/10.21741/9781644900390-8

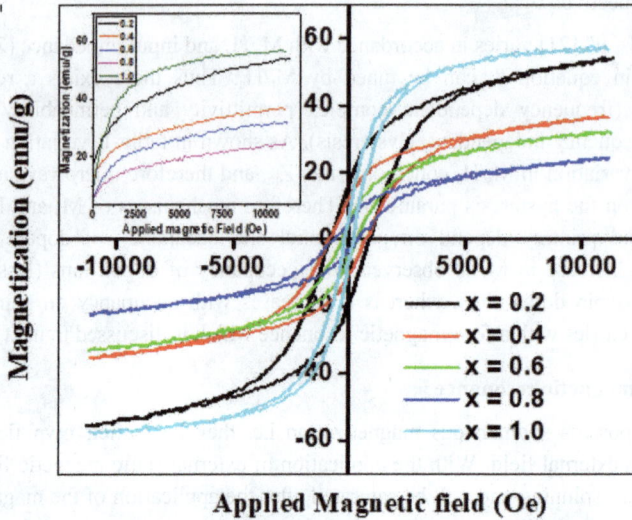

Fig. 5 *Magnetic hysteresis loops of $Ba_{0.5}Sr_{0.5}Co_xW_xFe_{12-2x}O_{19}$ (x = 0.2, 0.4, 0.6, 0.8 and 1.0) ferrite at room temperature. (Rights and Permission Elsevier) [23].*

Table 1. *Hysteresis parameters of M-type $Ba_{0.5}Sr_{0.5}Co_xGa_xFe_{12-2x}O_{19}$ ferrite (Rights and Permission Springer-Nature) [21].*

x	H_c	H_a	M_s	M_s/H_a (emu g^{-1}	F_r
	(Oe)	(kOe)	(emu/g)	/KOe)	(GHz)
0.0	250	9.4	55.8	5.94	4.19
0.2	800	9.32	59.6	6.39	4.16
0.4	150	6.86	48.2	7.03	3.06
0.6	200	6.74	62.2	9.23	3.01
0.8	80	4.78	53.6	11.21	2.13
1.0	60	4.48	56.1	12.52	2.0

Eng. Magnetic, Dielectric and Microwave Properties of Ceramics and Alloys   Materials Research Forum LLC
Materials Research Foundations **57** (2019) 175-190   doi: https://doi.org/10.21741/9781644900390-8

Thus microwave absorption or f_r can be tuned at the desired frequency by tuning the H_a. The dependence of H_a lies on the occupancy of doped ions on $4f_2$ and $2b$ sites [25], so preferential site occupancy of dopants is the prime factor in deciding the H_a. Nonetheless, saturation magnetization also (M_s) depends on the preferential site occupancy: it increases with dopants occupancy on spin down sites and decreases with spin up sites [26]. Therefore the selection of doped ions, along with preferential site occupancy, is very important to achieve absorption at the specific range of frequencies in the microwave region.

For Table 2, the absorption bandwidth of -10 dB and -20 dB in Co-Ga doped corresponds to that range/band of frequencies in which RL is more than > -10 dB and -20 dB respectively: this data can be derived from RL dips seen in figure 1. Composition x = 0.2 has -10 dB absorption bandwidth of 3.91 GHz and -20 dB bandwidth of 130 MHz. Therefore, wideband and narrowband absorber can be tailormade contingent on the application. Similarly, bandwidth of x=0.2 for other thickness is also shown.

Table 2 Different parameters for RL dips > -10 dB and -20 dB in Co-Ga doped M-type Ba-Sr hexaferrite (Rights and Permission Springer-Nature) [21]

x	Max. RL (dB)	Matching Thickness (mm)	Matching Frequency (GHz)	Frequency Band RL> -10 dB (GHz)	-10 dB Absorption Bandwidth (GHz)	Frequency Band RL> -20 dB (GHz)	-20 dB Absorption Bandwidth (GHz)
	-21.29	1.6	10.82	8.49-12.40	3.91	10.62-10.99	0.37
0.2	-23.92	1.7	9.98	8.20-11.88	3.68	9.65-10.31	0.66
	-26.13	1.8	9.79	-	-	9.11-10.02	0.91
	-23.64	1.9	9.03	-	-	8.32-9.22	0.90
	-29.74	2.0	8.28	-	-	8.20-9.01	0.81
	-23.10	2.1	8.20	-	-	8.20-8.33	0.13
	-12.48	2.3	8.20	8.20-9.0	0.80	-	-
	-10.08	2.4	8.20	8.20-8.23	0.03	-	-

Conclusions

We have discussed all types of dielectric relaxations which can be found in dielectrics or magnetodielectric ceramics. The role of eddy current, ferromagnetic resonance and static hysteresis loss in increasing the magnetic loss is explained. The necessity of low coercivity for a good absorber and role of other hysteresis parameters for maximal tunable absorption is mentioned. The optimization of microwave absorption by the quarterwavelength mechanism and, impedance matching accompanied by loss tangent and impedance matching mechanism is also described. The microwave absorption can be enhanced and tuned at the desired frequency and thickness in any ceramics through the different discussed mechanisms.

References

[1]   J. Tak, E. Jeong and J. Choi, "Metamaterial absorbers for 24-GHz automotive radar applications", Journal of Electromagnetic Waves and Applications, 31(6) (2017) 577-593. https://doi.org/10.1080/09205071.2017.1297257

[2]   Y. J. Yoo, J. S. Hwang and Y. P. Lee, "Flexible perfect metamaterial absorbers for electromagnetic wave", Journal of Electromagnetic Waves and Applications, 31(7) (2017) 663-715. https://doi.org/10.1080/09205071.2017.1312557

[3]   X. X. Xu, J. J. Jiang ,S.W. Bie,Q. Chen,C. K. Zhang and L. Miao, "Optimal Design Of Electromagnetic AbsorbersUsing Visualization Method For Wideband Potential Applications", Journal of Electromagnetic Waves and Applications, 26 (8-9) (2012) 1215-1225. https://doi.org/10.1080/09205071.2012.710575

[4]   Q. Deng, C. E. Huang, H. Wang, L. Zhao, C. Shen, "Microwave Dielectric Properties of $(1-x)(Ca_{0.88}Sr_{0.12})TiO_{3-x}(Bi_{0.5}Na_{0.5})TiO_3$ High Dielectric Constant Ceramics", Journal of Materials Science: Materials in Electronics, 29(5) (2018) 4035–4040. https://doi.org/10.1007/s10854-017-8346-8

[5]   J. Zhang, "Interference Effects on Microwave Absorbing Properties of W-Type BaZn2Fe16O27 Prepared by Solid Method", Journal of Materials Science: Materials in Electronics (2019). https://doi.org/10.1007/s10854-019-01162-x

[6]   Y. Liu, X. Su, F. Luo, J. Xu, J. Wang, X. He, Y. Qu, "Enhanced Electromagnetic and Microwave Absorption Properties of Hybrid $Ti_3SiC_2/BaFe_{12}O_{19}$ Powders", Journal of Electronic Materials, 48(4) (2019) 2364–2372. https://doi.org/10.1007/s11664-019-06928-x

[7]   U. J. Mahanta, M. Borah, N. S. Bhattacharyya, J. P. Gogoi, "High-Performance Broadband Microwave Absorbers Using Multilayer Dual-Phase Dielectric Composites", Journal of Electronic Materials, 48(4) (2019) 2438–2448. https://doi.org/10.1007/s11664-019-07038-4

Eng. Magnetic, Dielectric and Microwave Properties of Ceramics and Alloys   Materials Research Forum LLC
Materials Research Foundations **57** (2019) 175-190   doi: https://doi.org/10.21741/9781644900390-8

[8]   K. G. Kjelgard, D. T. Wisland and T. S. Lande, "3D Printed Wideband Microwave Absorbers using Composite Graphite/PLA Filament", *European Microwave Conference (EuMC),* Madrid, *48* (2018) 859-862. https://doi.org/10.23919/EuMC.2018.8541699

[9]   S. Ishiyama and N. Kuga, "Non-contact PIM measurement of dielectric wave absorbers by using a metallic resonator", *IEEE MTT-S International Microwave Symposium (IMS),* Honolulu, HI, (2017) 1270-1273. https://doi.org/10.1109/MWSYM.2017.8058841

[10]   R. Deng, K. Zhang, M. Li, L. Song, T. Zhang, "Targeted Design, Analysis and Experimental Characterization of Flexible Microwave Absorber for Window Application", Materials and Design, 162 (2019) 119–129. https://doi.org/10.1016/j.matdes.2018.11.038

[11]   H. Liao, L. Da, Z. Chen L. Tong, "Microporous Co/RGO Nanocomposites: Strong and Broadband Microwave Absorber with Well-Matched Dielectric and Magnetic Loss", Journal of Alloys and Compounds, 782 (2019) 556–565. https://doi.org/10.1016/j.jallcom.2018.12.241

[12]   J. Kuang, P. Jiang, X. Hou, T. Xiao, Q. Zheng, Q. Wang, W. Liu, W. Cao, "Dielectric Permittivity and Microwave Absorption Properties of SiC Nanowires with Different Lengths", Solid State Sciences, 91(2019) 73–76. https://doi.org/10.1016/j.solidstatesciences.2019.03.015

[13]   M. Ma, R. Yang, C. Zhang, B. Wang, Z. Zhao, W. Hu, Z. Liu, D. Yu, F. Wen, J. He, Y. Tian, "Direct Large-Scale Fabrication of C Encapsulated B 4 C Nanoparticles with Tunable Dielectric Properties as Excellent Microwave Absorbers." Carbon, 148 (2019) 504–511. https://doi.org/10.1016/j.carbon.2019.04.020

[14]   Y. Xu, J. Li, H. Ji, X. Zou, J. Zhang, Y. Yan, "Constructing Excellent Electromagnetic Wave Absorber with Dielectric-Dielectric Media Based on 3D Reduced Graphene and Ag(I)-Schiff Base Coordination Compounds", Journal of Alloys and Compounds, 781 (2019) 560–570. https://doi.org/10.1016/j.jallcom.2018.12.069

[15]   D. Zhang, Y. Ma, L. Jiang, X. Zhang, M. Yan, "Milimeter-Scale Metacomposite Absorbers by Structuring Ni@C Nanocapsules for Tunable Microwave Absorption", Journal of Alloys and Compounds, 784 (2019) 1205–1211. https://doi.org/10.1016/j.jallcom.2019.01.089

[16]   A. Ling, J. Pan, G. Tan, X. Gu, Y. Lou, S. Chen, Q. Man, R.-Wei Li, X. Liu, "Thin and Broadband $Ce_2Fe_{17}N_{3-\delta}$/MWCNTs Composite Absorber with Efficient Microwave Absorption", Journal of Alloys and Compounds, 787 (2019) 1097–1103. https://doi.org/10.1016/j.jallcom.2019.02.164

Eng. Magnetic, Dielectric and Microwave Properties of Ceramics and Alloys   Materials Research Forum LLC
Materials Research Foundations **57** (2019) 175-190   doi: https://doi.org/10.21741/9781644900390-8

[17]   H. Lv, G. Ji, H. Zhang, M. Li, Z. Zuo, Y. Zhao, B. Zhang, D. Tang and Y. Du, "CoxFey@C Composites with Tunable Atomic Ratios for Excellent Electromagnetic Absorption Properties," Sci. Rep., 5 (2015)18249 (1-5). https://doi.org/10.1038/srep18249

[18]   Y. Du, W. Liu, R. Qiang, Y. Wang, X. Han, J. Ma and P. Xu, "Shell Thickness-Dependent Microwave Absorption of Core–Shell Fe_3O_4@C Composites," ACS Appl. Mater .Interfaces, vol.6, pp. 12997, 2014. https://doi.org/10.1021/am502910d

[19]   B. Wang, J. Wei, Y. Yang, T. Wang and F. Li, "Investigation on peak frequency of the microwave absorption for carbonyl iron/epoxy resin composite," J. Magn. Magn. Mater., vol. 323, pp. 1101–1103, 2011. https://doi.org/10.1016/j.jmmm.2010.12.028

[20]   T. Inui, K. Konishi and K. Oda, "Fabrications of broad-band RF-absorber composed of planar hexagonal ferrites", IEEE Trans. Magn., 35(1999) 3148–3150. https://doi.org/10.1109/20.801110

[21]   H. Kaur, C. Singh, A. Marwaha, S. B. Narang, R. Jotania, S. R. Mishra, Y. Bai, K. C. James Raju, D. Singh, M. Ghimire, P. Dhruv, S. Sombra, "Elucidation of microwave absorption mechanisms in Co-Ga substituted Ba-Sr hexaferrites in X-band", Journal of Materials Science: Materials in Electronics, 29 (2018) 14995-15005. https://doi.org/10.1007/s10854-018-9638-3

[22]   G-Mei Shi, L. Sun, X. Wang, X. Bao, "Excellent electromagnetic wave absorption properties of LaOCl/C/MnO composites", Journal of Materials Science: Materials in Electronics 29 (2018) 2236-2243. https://doi.org/10.1007/s10854-017-8138-1

[23]   R. Joshi, C. Singh, D. Kaur, S. B. Narang, R. Jotania, J. Singh, Microwave absorption characteristics of Co^{2+} and W^{4+} substituted M-type $Ba_{0.5}Sr_{0.5}Co_xW_xFeO_{19}$ hexagonal ferrites, Journal of Materials Science: Materials in Electronics, 28 (2017) 228-235. https://doi.org/10.1007/s10854-016-5515-0

[24]   K.-K. Ji, Y. Li, M.-S. Cao, "Mn, Ti substituted barium ferrite to tune electromagnetic properties and enhanced microwave absorption", J. Material Science Materials in Electronics, 27 (2016) 5128-5135. https://doi.org/10.1007/s10854-016-4404-x

[25]   G. Mendoza-Suarez, L. P. Rivas-Vazquez, J. C. Corral-Huacuz, A. F. Fuentes, J. I. Escalante-Garcia, "Magnetic properties and microstructure of $BaFe_{11.6-2x}Ti_xM_xO_{19}$ (M = Co, Zn, Sn) compounds", Phy. B 339 (2003) 110–118. https://doi.org/10.1016/j.physb.2003.08.120

[26]   A. Ghasemi, A. Morisako, "Static and high frequency magnetic properties of Mn–Co–Zr substituted Ba-ferrite", J. Alloys Compd. 456 (2008) 485-491. https://doi.org/10.1016/j.jallcom.2007.02.101

Keyword Index

About the Editor

Dr. Charanjeet Singh is a Professor of Electronics and Communication at Lovely Professional University Phagwara Punjab India. His research area (15 years including teaching) encompasses tunable microwave ferrite absorbers/DRA/filters in GHz region, static V-I characteristics, sensors. Dr. Charanjeet Singh received his Ph.D degree and Bachelor of Technology (Electronics and Communication Engineering) from Guru Nanak Dev University, Amritsar Punjab in 2008 and 1996 respectively, Master Degree in the same discipline from Punjab Engineering College, Chandigarh in 2001. He has guided 3 Ph.D students and 6 Master students in the aforementioned fields and published 43 SCI papers in refereed journals with Scopus h-index 13 and 505 Scopus Citations. He is a recipient of the Young Scientist Award from International Union of Radio Science (URSI) at the conference held in Toyama, Japan in 2010. He has received the outstanding reviewer award from Elsevier Journal of Alloys and Compounds in 2017 and best teacher award from Indian Society for Technical Education. He has received travel grants from Govt. organizations like Department of Science and Technology (DST), Indian National Science Academy (INSA) and All India Council for Technical Education (AICTE) for attending conferences in Spain (2015), France (2012) and Turkey (2011). He has research collaborations with Dr. Yang Bai: Beijing University China, Dr. Sanjay Mishra: University of Memphis USA, Dr. Maceij Jaorszweski: Wroclaw University Poland, Dr. Ebtesam Ateia: Cairo University, Dr. Kohout Jaroslav: Charles University Czech Republic and Dr. Dariusz Bochenek: Silesia University Poland.

www.ingramcontent.com/pod-product-compliance
Lightning Source LLC
Chambersburg PA
CBHW071214210326
41597CB00016B/1811